Concrete-Filled Stainless Steel Tubular Columns

T0203659

Concrete-Filled Stainless Steel Tubular Columns

Vipulkumar Ishvarbhai Patel
Qing Quan Liang
Muhammad N. S. Hadi

CRC Press
Taylor & Francis Group
Boca Raton London New York

CRC Press is an imprint of the
Taylor & Francis Group, an **informa** business

CRC Press
Taylor & Francis Group
6000 Broken Sound Parkway NW, Suite 300
Boca Raton, FL 33487-2742

First issued in paperback 2020

ISBN-13: 978-1-138-54366-9 (hbk)
ISBN-13: 978-0-367-65683-6 (pbk)

Library of Congress Cataloging-in-Publication Data

Names: Patel, Vipulkumar, author. | Liang, Qing Quan, 1965- author. | Hadi, Muhammad N. S., author.
Title: Concrete-filled stainless steel tubular columns / Vipulkumar Patel, Qing Quan Liang and Muhammad Hadi.
Description: Boca Raton: Taylor & Francis, CRC Press, [2019] | Includes bibliographical references and index. |
Identifiers: LCCN 2018042188 (print) | LCCN 2018042497 (ebook) | ISBN 9781351005685 (ePub) | ISBN 9781351005692 (Adobe PDF) | ISBN 9781351005678 (Mobipocket) | ISBN 9781138543669 (hardback) | ISBN 9781351005708 (ebook)
Subjects: LCSH: Columns, Concrete. | Concrete-filled tubes. | Structural analysis (Engineering) | Tubes, Steel.
Classification: LCC TA683.5.C7 (ebook) | LCC TA683.5.C7 P38 2019 (print) | DDC 624.1/7725—dc23
LC record available at https://lccn.loc.gov/2018042188

**Visit the Taylor & Francis Web site at
http://www.taylorandfrancis.com**

**and the CRC Press Web site at
http://www.crcpress.com**

Contents

Preface

Concrete-filled stainless steel tubular (CFSST) columns are increasingly used in modern composite construction due to their distinguished features, such as high strength, high ductility, aesthetic appearance, high corrosion resistance, high durability, and ease of maintenance. Thin-walled CFSST columns are characterized by the different strain-hardening behaviors of stainless steel in tension and in compression, local buckling of stainless steel tubes, and concrete confinement. The current design codes and existing numerical models that do not account for these characteristics may either overestimate or underestimate the ultimate strengths of CFSST columns.

This book is the first monograph on the nonlinear analysis, behavior, and design of CFSST columns. It presents accurate and efficient computational models based on the fiber element method for predicting the behavior of circular and rectangular CFSST short and slender columns under axial load and biaxial bending. The effects of different strain-hardening characteristics of stainless steel in tension and in compression, progressive local buckling of rectangular stainless steel tubes, concrete confinement, and geometric and material nonlinearities are taken into consideration in the computational models. The mathematical models accurately simulate the axial load–strain behavior, moment–curvature curves, axial load–deflection responses, and axial load–moment strength interaction diagrams of CFSST columns. This book describes the formulations of mathematical models, computational procedures, model verifications, behavior, and design of circular and rectangular CFSST short and slender columns.

This book is written for practicing structural and civil engineers, academic researchers, and undergraduate and postgraduate students in civil engineering who are interested in the latest computational technologies and design methods for CFSST columns.

Chapter 1 introduces the composite construction of stainless steel and concrete and the material properties of stainless steel grades. The nonlinear analysis, behavior, and design of circular and rectangular CFSST short columns under axial compression, combined axial load and bending. or biaxial loads are presented in Chapter 2. Chapter 3 describes mathematical models and modeling procedures for predicting the axial load–deflection responses and axial load–moment strength envelopes of circular CFSST slender beam-columns subjected to eccentric loading. The behavior and design of circular CFSST slender beam-columns are also discussed. The nonlinear analysis of rectangular CFSST slender beam-columns under combined axial load and biaxial bending is treated in Chapter 4. This chapter covers the formulations of theoretical models and computer simulation procedures for CFSST columns under biaxial loads, and the fundamental behavior and design of rectangular CFSST slender columns in accordance with Eurocode 4.

Vipulkumar Ishvarbhai Patel
Qing Quan Liang
Muhammad N. S. Hadi
Melbourne, VIC, Australia

Acknowledgments

The authors would like to thank Distinguished Professor Yi-Min Xie at RMIT University, Emeritus Professor Grant P. Steven at the University of Sydney and Strand7 Pty Ltd, Professor Brian Uy at the University of Sydney, Laureate Professor Mark A. Bradford at the University of New South Wales, Professor Jat-Yuen Richard Liew at the National University of Singapore, Professor Yeong-Bin Yang at National Taiwan University and Chongqing University, Professor Yanglin Gong at Lakehead University, Dr. Sawekchai Tangaramvong at Chulalongkorn University, and Emeritus Professor N. E. Shanmugam at Anna University for their useful communications and support. Finally, the authors thank their families for their great encouragement and support.

Authors

Vipulkumar Ishvarbhai Patel is a Lecturer of Structural Engineering at La Trobe University, Australia, and the co-author of *Nonlinear Analysis of Concrete-Filled Steel Tubular Columns* (2015, Scholars' Press).

Qing Quan Liang is an Associate Professor of Structural Engineering at Victoria University, Australia, the Founder and President of Australian Association for Steel-Concrete Composite Structures (AASCCS), the author of *Performance-Based Optimization of Structures: Theory and Applications* (2004, Spon Press) and *Analysis and Design of Steel and Composite Structures* (2014, CRC Press), and the co-author of *Nonlinear Analysis of Concrete-Filled Steel Tubular Columns* (2015, Scholars' Press).

Muhammad N. S. Hadi is an Associate Professor of Structural Engineering at the University of Wollongong, Australia, and the co-author of *Nonlinear Analysis of Concrete-Filled Steel Tubular Columns* (2015, Scholars' Press) and *Earthquake Resistant Design of Buildings* (2017, CRC Press).

Chapter 1

Introduction

1.1 BACKGROUND

Stainless steels have distinguished advantages over carbon steels, including aesthetic appearance, corrosion resistance, ductility, fire resistance, durability, energy absorption capacity, and ease of maintenance. Stainless steels have traditionally been used in minor building components and façades because of their aesthetic appearance and excellent corrosion resistance – for example, in the Chrysler Building in New York, completed in 1930, and the Lloyd's Building in London, completed in 1986 (Mann 1993; Baddoo 2008; Gardner 2008; Gedge 2008). The initial cost of stainless steels is about four times that of carbon steels, which has greatly limited the widespread applications of stainless steels in structural engineering. However, the life cycle cost of stainless steel structures, which includes initial material costs, costs associated with initial corrosion, fire protection costs, maintenance costs, and end-of-life costs, is much lower than that of carbon steel structures (Gardner et al. 2007). Climate changes have demanded the design and construction of sustainable buildings and infrastructure with the least materials, low life cycle costs, and minimum environmental impacts. To meet this challenging demand, stainless steels have increasingly been used as structural members in buildings, bridges, offshore structures, and nuclear power plants (Baddoo 2008). Some of the practical engineering examples are the Grande Arche de la Défense in Paris, completed in 1989; the Sanomatalo building in Helsinki, Finland, completed in 1999; the road bridge in Menorca in Spain; and the Millennium Footbridge in the City of York, England, completed in 2001 (Gardner 2008).

To reduce the initial high cost of stainless steel hollow columns, thinner stainless steel elements can be used to form the hollow columns. However, thin stainless steel tubular columns may undergo local buckling, which reduces their ultimate strengths. The economical utilization of stainless steel material can be achieved by means of filling concrete into hollow stainless steel tubular columns to make them concrete-filled stainless steel tubular (CFSST) columns. The filled concrete delays the local buckling of the stainless steel tube, forces the tube to buckle locally outward, and significantly increases the ultimate strength of the hollow stainless steel column. In addition, the concrete infill significantly improves the fire resistance of the hollow stainless steel tubular column. Further economies can be achieved by filling high-strength concrete into thin-walled stainless steel tubular columns (Mann 1993; Gardner 2005; Baddoo 2008; Han et al. in press). Figure 1.1 depicts the cross-sections of the most commonly used CFSST columns. They can be classified into two groups: CFSST columns (Figure 1.1a–d) and concrete-filled double steel tubular (CFDST) columns with outer stainless steel jacket, as shown in Figure 1.1e–g (Han et al. 2011, in press).

Extensive experimental and numerical studies on normal and high-strength concrete-filled steel tubular (CFST) columns made of carbon steel tubes have been carried out by researchers (O'Shea and Bridge 2000; Uy 2000, 2001; Sakino et al. 2004; Giakoumelis and Lam 2004; Liang 2009a, b; Liang 2011a, b; Tao et al. 2013; Han et al. 2014; Liew et al. 2016; Liang 2017, 2018; Kamil et al. 2018; Ahmed et al. 2018). The behavior of CFST columns has been well understood from these studies. It should be noted that the material behavior of stainless steels is significantly different from that of carbon steels. However, there have been relatively limited investigations on the behavior of CFSST columns under various loading conditions (Young and Ellobody 2006; Lam and Gardner 2008; Tao et al. 2011; Uy et al. 2011; Hassanein et al. 2013; Patel et al. 2017; Han et al. in press). To reduce the initial material cost of stainless steel, Ye et al. (2016) and Ye et al. (2018) proposed concrete-filled bimetallic steel tubular (CFBST) columns, in which the cross-section consists of an external stainless steel thin layer and an inner carbon steel tube as shown in Figure 1.1c, d. The use of the thin stainless steel layer to cover the CFST column significantly reduces the amount of stainless steel used in the column. Consequently, this results in a significant reduction in material costs.

Due to the lack of sufficient numerical and experimental investigations on the responses of CFSST short and slender columns under axial load and bending, current design codes, such as Eurocode 4 (2004) and AISC 360-16 (2016), do not provide design rules for the design of CFSST columns. Recently, Liang (2014) and Patel et al. (2015) have published books on the nonlinear analysis and design of composite columns made of carbon steels. However, no research books on CFSST columns are available. This book is the first monograph on the nonlinear analysis, behavior, and design of CFSST columns. It presents accurate and efficient computational models for determining the nonlinear inelastic responses of circular and rectangular CFSST short and slender columns under various loading conditions, including axial compression, combined axial compression and uniaxial bending, and combined axial compression and biaxial bending. The numerical methods presented in this book can be extended to other types of composite columns. The following important features are described in this book:

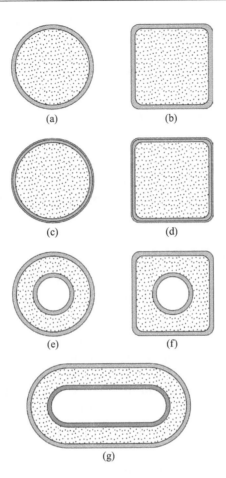

Figure 1.1 Typical cross-sections of composite columns: (a) circular CFSST columns; (b) square CFSST columns; (c) circular CFBST columns; (d) rectangular CFBST columns; (e) circular CFDST columns with outer stainless steel tube; (f) rectangular CFDST columns with inner circular carbon steel tube; (g) round-ended rectangular CFDST columns.

- Various stress–strain models proposed by researchers for stainless steels;
- Fiber-based mathematical models for the nonlinear analysis of rectangular and circular CFSST short and slender beam-columns under various loading conditions;
- The modeling strategies to incorporate the effects of progressive local buckling on the performance of thin-walled rectangular CFSST beam-columns;
- Computer simulation procedures for calculating the axial load–strain responses, axial load–deflection responses, and axial load–moment strength envelopes of CFSST columns;

- The fundamental behavior of CFSST short and slender columns with important geometric and material parameters; and
- The design of CFSST short and slender columns using the current design codes for CFST columns and equations proposed by researchers.

1.2 STAINLESS STEEL GRADES

Stainless steels are generally classified into five grades based on their microstructure, which depends on the chemical composition (AS/NZS 4673:2001 2001). Each grade has different material properties in terms of strength, resistance to corrosion, and ease of fabrication. These five stainless steel grades are introduced in this section.

1.2.1 Austenitic stainless steels

Austenitic stainless steels are the most commonly used stainless steels in engineering structures, designated as grades 304, 316, 304L, and 316L. The grades 304L and 316L are the low carbon variants of grades 304 and 316. These stainless steels contain 18% chromium and 8% nickel in their chemical composition. The chromium content of the grade 316 austenitic stainless steel is slightly lower than that of the grade 304 one, but grade 316 stainless steel contains an additional 2% molybdenum to increase its resistance against corrosion. Grades 304L and 316L austenitic stainless steels have a maximum of 0.03% carbon, which greatly reduces the residual stresses induced from the welding process. Austenitic stainless steels have good corrosion resistance and high ductility, are easily fabricated in different shapes, and readily weldable. These stainless steels become stronger by cold working but not by the heat treatment. The mechanical properties of austenitic stainless steels are given in Tables 1.1 and 1.2. It can be seen from Tables 1.1 and 1.2 that the strengths of austenitic stainless steels are comparable to carbon steels.

1.2.2 Ferritic stainless steels

Ferritic stainless steels with a ferritic microstructure have little nickel content in their chemical composition. The cost of ferritic stainless steels is much lower than that of austenitic stainless steels due to the little nickel content. Ferritic stainless steels have lower weldability, ductility, and formability than austenitic stainless steels. Despite the corrosion resistance of ferritic grades being lower than that of austenitic ones, their stress corrosion cracking resistance is higher. The ferritic stainless steel can be made stronger by cold working, not by heat treatment. However, the strength increased by the cold working process for ferritic grades is much less than for austenitic stainless steels. Tables 1.1 and 1.2 provide the material properties of ferritic grades EN 10088 1.4003, 409, and 430. It appears that

Table 1.1 Material properties of stainless steels for longitudinal tension

Property	Stainless steel grade					
	304, 316	304L, 316L	409	1.4003	430	S31803
Initial elastic modulus E_0 (GPa)	195	195	185	195	185	200
0.2% proof stress $\sigma_{0.2}$ (MPa)	205	205	205	280	275	430
Ramberg–Osgood parameter n	7.5	7.5	11	9	8.5	5.5
Ultimate strength (MPa)	520	485	380	435	450	590

Source: Adapted from AS/NZS 4673:2001. Australian/New Zealand Standard for cold-formed stainless steel structures. Standards Australia, Sydney, NSW, Australia, 2001.

Table 1.2 Material properties of stainless steels for longitudinal compression

Property	Stainless steel grade					
	304, 316	304L, 316L	409	1.4003	430	S31803
Initial elastic modulus E_0 (GPa)	195	195	185	210	185	195
0.2% proof stress $\sigma_{0.2}$ (MPa)	195	195	205	260	275	435
Ramberg–Osgood parameter n	4	4	9.5	7.5	6.5	5.0

Source: Adapted from AS/NZS 4673:2001. Australian/New Zealand Standard for cold-formed stainless steel structures. Standards Australia, Sydney, NSW, Australia, 2001.

the strength of ferritic grades in annealed condition is similar to that of austenitic stainless steels.

1.2.3 Martensitic stainless steels

Unlike austenitic and ferritic grades, heat treatments can be used to harden martensitic stainless steels to increase their strengths. However, the toughness of martensitic stainless steels is not strong enough for their use in structural members. The martensitic stainless steels cannot be utilized in welded connections, but they can be used in bolted connections.

1.2.4 Duplex stainless steels

The duplex stainless steels are formed by a mixed microstructure of austenitic and ferritic stainless steels so that they have the best properties of austenitic and ferritic grades. The 0.2% proof stress and corrosion resistance of duplex stainless steels are higher than those of austenitic grades. However, the weldability and formability of duplex stainless steels are lower than those of austenitic ones. The duplex stainless steels can be hardened by cold working. The UNS S31803 grade is the most commonly used duplex stainless steel in engineering structures. The chemical composition of duplex stainless steels has 22% chromium, 5% nickel, and 3% molybdenum. The mechanical properties of duplex stainless steels are listed in Tables 1.1 and 1.2.

1.2.5 Precipitation hardening stainless steels

Precipitation hardening stainless steels have the highest strengths, which are obtained by means of heat treatment in comparison with all other stainless steel grades. This stainless steel grade can have 0.2% proof stress greater than 1000 MPa. However, precipitation-hardening stainless steels are not suitable for use in welded connections. This is because heat treatment and surface finishing of these steels after welding are required. The UNS A17400 called grade 630 is the most commonly used precipitation hardening stainless steel.

1.3 BASIC STRESS–STRAIN BEHAVIOR OF STAINLESS STEELS

The stress–strain behavior of stainless steels is significantly different from that of carbon steels. Figure 1.2 illustrates the typical stress–strain responses of stainless steels and carbon steels. It can be seen from Figure 1.2 that the shape of the stress–strain curve for stainless steels is significantly different from that for carbon steels. The carbon steel generally follows the linear elastic stress–strain response up to the yield strength and a horizontal plateau before attaining strain-hardening. In contrast to the carbon steel, stainless steel exhibits the rounded stress–strain relationship without a well-defined yield stress. Therefore, the 0.2% proof stress is usually utilized to represent the yield strength of stainless steel. The roundness of the nonlinear stress–strain curve for stainless steel generally depends on the chemical composition, cold working, and heat treatment. In addition, stainless steel exhibits quite different strain-hardening behaviors in tension and compression. The compressive ultimate strength of stainless steel is sustainably higher than that of the tensile ultimate strength. Moreover, stainless steels have superior ductility.

1.4 CHARACTERISTICS OF CFSST COLUMNS

1.4.1 Concrete confinement in circular CFSST columns

The stainless steel tube offers confinement to the concrete in a circular CFSST column under axial compression (Liang 2014; Patel et al. 2015). Both the compressive strength and ductility of the filled concrete are

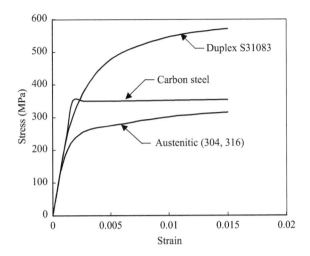

Figure 1.2 Comparison of stress–strain curves for stainless steels and carbon steels. (Source: Adapted from AS/NZS 4673:2001. Australian/New Zealand Standard for cold-formed stainless steel structures. Standards Australia, Sydney, NSW, Australia, 2001.)

remarkably increased by the confinement effect. As a result of this, the ultimate axial strength and ductility of the circular CFSST column are found to increase. The confinement mechanism in a circular CFSST column under increasing axial compression is developed by the interaction between the circular stainless steel tube and the concrete infill. The stainless steel tube and the filled concrete are in the elastic range at the early stage of loading. The Poisson's ratio of the elastic stainless steel material is larger than that of the elastic concrete material. The stainless steel tube is subjected to hoop compressive stresses while the concrete exhibits lateral tension. At this stage of loading, the concrete core is not confined by the circular stainless steel tube. As the axial compressive load increases and the concrete compressive stress is greater than its unconfined compressive strength, the concrete expanses laterally more than the stainless steel tube due to the effect of the Poisson's ratio. There is a radial pressure (f_{rp}) at the interface between the concrete core and the stainless steel tube as illustrated in Figure 1.3. The stainless steel tube is subjected to longitudinal compressive stress and hoop tensile stress (f_{rs}), while the concrete core is under triaxial compressive stresses. The concrete core is therefore confined by the circular stainless steel tube. The hoop tension reduces the 0.2% proof stress of the stainless steel tube. The effects of concrete confinement on the behavior of circular CFSST columns are considered in the computational models presented in this book.

Figure 1.4 illustrates the axial load–strain responses of stainless steel and concrete components in a circular short CFSST column under axial

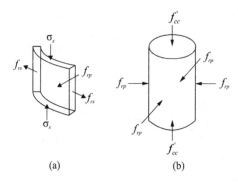

Figure 1.3 Radial pressure in a circular CFSST column: (a) radial pressures on the element of stainless steel tube; (b) radial confining pressure on the concrete core.

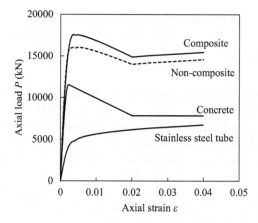

Figure 1.4 Comparison of component strengths in a circular CFSST column.

compression. The diameter of the stainless steel tube was 600 mm with a diameter-to-thickness ratio of 100. The 0.2% proof stress of the stainless steel was 430 MPa and the nonlinearity index was 5.5. The stainless steel tube was filled with 50 MPa concrete. As shown in the figure, the sum of the concrete and steel component strengths as a non-composite column is much higher than that of the individual component alone. It should be noted that the composite action reflected by the confinement effect increases the ultimate axial strength of the CFSST column by 10% in comparison with the non-composite one. The CFSST column has higher initial stiffness than the steel and concrete component alone. This clearly indicates the benefit achieved by the composite action in terms of the confinement mechanism in circular CFSST columns (Patel et al. 2015).

1.4.2 Local buckling of rectangular CFSST columns

A rectangular thin-walled hollow stainless steel tubular short column subjected to axial compression may fail by the outward and inward local buckling of the stainless steel tube walls. This buckling mode is illustrated in Figure 1.5a. It appears that the two opposite tube walls buckle locally outward and the adjacent side walls buckle locally inward. In contrast, thin-walled rectangular CFSST short columns under axial compression may fail by the outward local buckling. Figure 1.5b depicts the local buckling mode of axially loaded thin-walled CFSST short columns. It is observed from the figure that the stainless steel tube of the CFSST column buckles locally outward. This is because the concrete core prevents the inward local buckling of the stainless steel tube. The restraint provided by the concrete delays the local buckling and significantly increases the post-local buckling strength of the stainless steel tube of the CFSST column. The local buckling of the stainless steel tube results in the transfer of column loads from the buckled stainless steel portions to the adjacent concrete. This may lead to the crushing of the concrete near the buckled stainless steel tube wall (Ge and Usami 1992; Uy 2000). The local buckling of thin-walled CFSST columns with large width-to-thickness ratios significantly reduces the column ultimate strengths and therefore must be taken into account in the

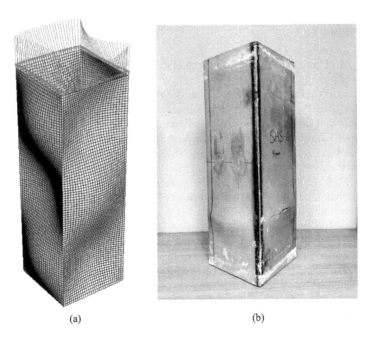

(a)   (b)

Figure 1.5 Local buckling modes: (a) hollow stainless steel tubular short column; (b) CFSST short column.

nonlinear inelastic numerical analysis of CFSST columns. The behavior of thin-walled rectangular slender CFSST columns is generally characterized by the overall column buckling. However, when non-compact or slender stainless steel sections are used to construct CFSST slender columns, the interaction local and overall buckling may occur. The effects of the interaction local and overall buckling on the responses of thin-walled rectangular CFSST slender columns are taken into account in the mathematical models presented in this book.

1.5 CONCLUSIONS

This chapter has presented an introduction to CFSST columns. The advantages of stainless steel materials for structural applications have been described. The material properties of five stainless steel grades have been discussed. The basic stress–strain behavior of stainless steels in comparison with carbon steels has been presented. The silent features of CFSST columns have been highlighted, including the concrete confinement in circular CFSST columns and the local buckling of thin-walled rectangular CFSST columns. These important features must be considered in the nonlinear analysis and design of short and slender CFSST columns under axial compression and combined action of axial compression and bending.

REFERENCES

Ahmed, M., Liang, Q. Q., Patel, V. I. and Hadi, M. N. S. (2018) "Nonlinear analysis of rectangular concrete-filled double steel tubular short columns incorporating local buckling," *Engineering Structures*, 175: 13–26.

AISC 360-16 (2016) *Specification for Structural Steel Buildings*, Chicago, IL, USA: American Institute of Steel Construction.

AS/NZS 4673:2001 (2001) *Australian/New Zealand Standard for Cold-Formed Stainless Steel Structures*, Sydney, NSW, Australia: Standards Australia and Standards New Zealand.

Baddoo, N. R. (2008) "Stainless steel in construction: A review of research, applications, challenges and opportunities," *Journal of Constructional Steel Research*, 64(11): 1199–1206.

Eurocode 4 (2004) *Design of Composite Steel and Concrete Structures, Part 1.1: General Rules and Rules for Building*, Brussels, Belgium: European Committee for Standardization, CEN.

Gardner, L. (2005) "The use of stainless steel in structures," *Progress in Structural Engineering and Materials*, 7: 45–55.

Gardner, L. (2008) "Aesthetics, economics and design of stainless steel structures," *Advanced Steel Construction*, 4(2): 113–122.

Gardner, L., Cruise, R. B., Sok, C. P., Krishnan, K. and Ministro, J. (2007) "Life cycle costing of metallic structures," *Proceedings of the Institution of Civil Engineers, Engineering Sustainability*, 160(4): 167–177.

Ge, H. B. and Usami, T. (1992) "Strength of concrete-filled thin-walled steel box columns: Experiment," *Journal of Structural Engineering*, ASCE, 118(11): 3036–3054.

Gedge, G. (2008) "Structural uses of stainless steel – buildings and civil engineering," *Journal of Constructional Steel Research*, 64(11): 1194–1198.

Giakoumelis, G. and Lam, D. (2004) "Axial capacity of circular concrete-filled tube columns," *Journal of Constructional Steel Research*, 60(7): 1049–1068.

Han, L. H., Li, W. and Bjorhovde, R. (2014) "Developments and advanced applications of concrete-filled steel tubular (CFST) structures: Members," *Journal of Constructional Steel Research*, 100: 211–228.

Han, L. H., Ren, Q. X. and Li, W. (2011) "Tests on stub stainless steel-concrete-carbon steel double-skin tubular (DST) columns," *Journal of Constructional Steel Research*, 67(3): 437–452.

Han, L. H., Xu, C. Y. Tao, Z. (in press) "Performance of concrete filled stainless steel tubular (CFSST) columns and joints: Summary of recent research," *Journal of Constructional Steel Research*. doi:10.1016/j.jcsr.2018.02.038

Hassanein, M. F., Kharoob, O. F. and Liang, Q. Q. (2013) "Circular concrete-filled double skin tubular short columns with external stainless steel tubes under axial compression," *Thin-Walled Structures*, 73: 252–263.

Kamil, G. M., Liang, Q. Q. and Hadi, M. N. S. (2018) "Local buckling of steel plates in concrete-filled steel tubular columns at elevated temperatures," *Engineering Structures*, 168: 108–118.

Lam, D. and Gardner, L. (2008) "Structural design of stainless steel concrete filled columns," *Journal of Constructional Steel Research*, 64(11): 1275–1282.

Liang, Q. Q. (2009a) "Performance-based analysis of concrete-filled steel tubular beam-columns, Part I: Theory and algorithms," *Journal of Constructional Steel Research*, 65(2): 363–372.

Liang, Q. Q. (2009b) "Performance-based analysis of concrete-filled steel tubular beam-columns, Part II: Verification and applications," *Journal of Constructional Steel Research*, 65(2): 351–362.

Liang, Q. Q. (2011a) "High strength circular concrete-filled steel tubular slender beam-columns, Part I: Numerical analysis," *Journal of Constructional Steel Research*, 67(2): 164–171.

Liang, Q. Q. (2011b) "High strength circular concrete-filled steel tubular slender beam-columns, Part II: Fundamental behavior," *Journal of Constructional Steel Research*, 67(2): 172–180.

Liang, Q. Q. (2014) *Analysis and Design of Steel and Composite Structures*, Boca Raton, FL, USA and London, UK: CRC Press, Taylor & Francis Group.

Liang, Q. Q. (2017) "Nonlinear analysis of circular double-skin concrete-filled steel tubular columns under axial compression," *Engineering Structures*, 131: 639–650.

Liang, Q. Q. (2018) "Numerical simulation of high strength circular double-skin concrete-filled steel tubular slender columns," *Engineering Structures*, 168: 205–217.

Liew, J. Y. R., Xiong, M. and Xiong, D. (2016) "Design of concrete filled tubular beam-columns with high strength steel and concrete," *Structures*, 8: 213–226.

Mann, A. P. (1993) "The structural use of stainless steel," *The Structural Engineer*, 71(4): 60–69.

O'Shea, M. D. and Bridge, R. Q. (2000) "Design of circular thin-walled concrete filled steel tubes," *Journal of Structural Engineering*, ASCE, 126(11): 1295–1303.

Patel, V. I., Liang, Q. Q. and Hadi, M. N. S. (2015) *Nonlinear Analysis of Concrete-Filled Steel Tubular Columns*, Germany: Scholar's Press.

Patel, V. I., Liang, Q. Q. and Hadi, M. N. S. (2017) "Nonlinear analysis of circular high strength concrete-filled stainless steel tubular slender beam-columns," *Engineering Structures*, 130: 1–13.

Sakino, K., Nakahara, H., Morino, S. and Nishiyama, I. (2004) "Behavior of centrally loaded concrete-filled steel-tube short columns," *Journal of Structural Engineering*, ASCE, 130(2): 180–188.

Tao, Z., Uy, B., Liao, F. Y. and Han, L. H. (2011) "Nonlinear analysis of concrete-filled square stainless steel stub columns under axial compression," *Journal of Constructional Steel Research*, 67(11): 1719–1732.

Tao, Z., Wang, Z. B. and Yu, Q. (2013) "Finite element modelling of concrete-filled steel stub columns under axial compression," *Journal of Constructional Steel Research*, 89: 121–131.

Uy, B. (2000) "Strength of concrete-filled steel box columns incorporating local buckling," *Journal of Structural Engineering*, ASCE, 126(3): 341–352.

Uy, B. (2001) "Strength of short concrete filled high strength steel box columns," *Journal of Constructional Steel Research*, 57(2): 113–134.

Uy, B., Tao, Z. and Han, L. H. (2011) "Behaviour of short and slender concrete-filled stainless steel tubular columns," *Journal of Constructional Steel Research*, 67(3): 360–378.

Ye, Y., Han, L. H., Sheehan, T. and Guo, Z. X. (2016) "Concrete-filled bimetallic tubes under axial compression: Experimental investigation," *Thin-Walled Structures*, 108: 321–332.

Ye, Y., Zhang, S. J., Han, L. H. and Liu, Y. (2018) "Square concrete-filled stainless steel/carbon steel bimetallic tubular stub columns under axial compression," *Journal of Constructional Steel Research*, 146: 49–62.

Young, B. and Ellobody, E. (2006) "Experimental investigation of concrete-filled cold-formed high strength stainless steel tube columns," *Journal of Constructional Steel Research*, 62(5): 484–492.

Chapter 2

Nonlinear analysis of CFSST short columns

2.1 INTRODUCTION

The accuracy of numerical models for the nonlinear inelastic analysis of concrete-filled stainless steel tubular (CFSST) columns relies on the implementation of accurate material stress–strain relationships for stainless steel and concrete. Several stress–strain models for stainless steels have been developed by researchers (Rasmussen 2003; Gardner and Nethercot 2004; Quach et al. 2008; Abdella et al. 2011; Tao and Rasmussen 2016). Rasmussen (2003) proposed two-stage stress–strain relationships for stainless steel based on tension coupon test data. However, the stress–strain model by Rasmussen (2003) does not recognize the different strain-hardening characteristics of stainless steel in tension and compression so that it underestimates the ultimate axial strengths of short CFSST columns as reported by Patel et al. (2014). Quach et al. (2008) analyzed experimental results on stainless steel under tension and compression and developed a three-stage stress–strain model for stainless steels accounting for different strain-hardening characteristics in tension and compression. Abdella et al. (2011) presented the inversion of the three-stage stress–strain model proposed by Quach et al. (2008) for stainless steel, which expresses the stress as a function of strain.

Experimental studies indicated that the confinement offered by the circular steel tube in a circular concrete-filled steel tubular (CFST) column increased both the strength and ductility of the filled concrete, but the confinement provided by the rectangular steel tube to the concrete improved only the ductility of the concrete in a rectangular CFST column (Klöppel and Goder 1957; Furlong 1967; Knowles and Park 1969; Tomii and Sakino 1979; Schneider 1998; O'Shea and Bridge 2000; Han 2002; Sakino et al. 2004; Giakoumelis and Lam 2004; Tao and Han 2006). The confinement is caused by the interaction between the steel tube and the filled concrete. The lateral expansion of the concrete core under the increasing load causes the expansion of the circular steel tube in the hoop direction, which develops hoop tension in the steel tube. The circular steel tube provides lateral confining pressures to the infilled concrete. Confinement models for determining the lateral confining pressures on concrete in circular CFST columns were proposed by Hu et al. (2003) and Liang and Fragomeni (2009), which are a function of the geometric and material properties of the CFST columns.

Thin-walled rectangular CFST columns made of non-compact or slender steel sections under applied loads are susceptible to the outward local buckling. Experiments indicated that the four steel tube walls of a rectangular CFST column under axial compression buckled locally outward (Ge and Usami 1992; Wright 1995; Bridge and O'Shea 1998; Uy 1998, 2000, 2001). The local buckling of the steel tube considerably reduces the capacity of thin-walled CFST columns. The steel tube may buckle before or after attaining the ultimate load, which depends on the depth-to-thickness ratio and material strengths of the rectangular CFST column. The finite element analyses were conducted by Liang and Uy (1998, 2000) and Liang et al. (2004, 2007a) to study the local and post-local buckling behavior of clamped steel plates in rectangular short CFST beam-columns subjected to axial load and biaxial bending and of double skin composite panels. They proposed equations for predicting the effective width and strength of steel plates in thin-walled CFST columns under axial load and biaxial bending. These equations can be incorporated in the fiber element analysis techniques to compute the initial local and post-local buckling strengths of steel plates in rectangular CFST columns (Liang et al. 2006, 2007b; Liang 2009a–c).

Experimental investigations on the behavior of CFST columns subjected to axial compression have been carried out by researchers (Schneider 1998; O'Shea and Bridge 2000; Sakino et al. 2004; Giakoumelis and Lam 2004). However, research studies on short CFSST beam-columns have relatively been limited. The experimental study on rectangular short CFSST columns under concentric compression was carried out by Young and Ellobody (2006). They reported that the use of material strengths obtained from the tensile couple tests underestimated the ultimate axial loads of short CFSST columns. This was because the enhanced strain-hardening of stainless steel in compression was not considered. Lam and Gardner (2008) conducted experiments on circular short CFSST columns under axial loading. They proposed design equations based on the Continuous Strength Method for estimating the axial capacity of circular short CFSST columns. An experimental program on short and slender CFSST columns was carried out by Uy et al. (2011). They reported that the current design provisions for conventional CFST columns provide the conservative ultimate axial loads of CFSST columns. Xing and Young (2018) presented experimental results on rectangular lean duplex short CFSST columns under concentric axial compression. Their study indicated that the current design codes yield conservative solutions to the ultimate axial strengths of rectangular lean duplex short CFSST columns.

The finite element analysis software Abaqus and fiber-based simulation techniques have been employed to study the responses of CFST columns (Hu et al. 2003; Ellobody et al. 2006; Liang et al. 2006; Liang et al. 2007b; Liang 2009a–c; Liang and Fragomeni 2009, 2010; Tao et al. 2013; Thai et al. 2014). The performance of axially loaded square short CFSST columns was numerically studied by Ellobody and Young (2006) employing the finite element software Abaqus. However, the significant strain-hardening of stainless steel in compression was not considered in their finite element model. Tao et al. (2011) utilized the finite element software Abaqus to model the inelastic cross-section behavior of square short CFSST columns under concentric loading. The two-stage stress–strain relations of stainless steel developed by Rasmussen (2003) was adopted in the analysis. The structural performance of circular lean duplex short CFSST columns under axial compression was numerically investigated by Hassanein et al. (2013). Patel et al. (2014) developed a fiber-based model for the predictions of the axial load–strain responses of circular short CFSST columns. The three-stage stress–strain laws of stainless steel given by Quach et al. (2008) and Abdella et al. (2011) were implemented in the fiber model that recognize different strain-hardening characteristics in tension and compression.

This chapter describes the nonlinear inelastic analysis of circular and rectangular CFSST short columns under axial compression or combined axial compression and bending. The nonlinear analysis technique is based on the fiber element method. Several stress–strain constitutive models for carbon steels and stainless steels proposed by researchers are given in detail. The material constitutive laws of confined concrete in circular and rectangular CFSST columns are discussed. The simulation of the progressive local and post-local buckling of rectangular stainless steel tubes is presented. The basic theory and computational procedures for determining the axial load–strain curves, moment–curvature curves, and axial load–moment interaction curves for CFSST short columns are provided. Comparative studies are undertaken to validate the effective width models and the fiber-based modeling technique, to evaluate various stress–strain models for stainless steels, and to compare the behavior of CFST and CFSST columns. The fundamental behavior and design of CFSST short columns are discussed.

2.2 STRESS–STRAIN RELATIONSHIPS OF CARBON STEELS

Mild, cold-formed, and high-strength steels are generally used in the construction of conventional CFST columns. The idealized stress–strain curves for structural steels are illustrated in Figure 2.1. The mild steel exhibits an elastic response and a plateau before attaining strain-hardening. Unlike mild steel, cold-formed steel follows the rounded stress–strain curve instead of a plateau. For high-strength steel, the rounded part of the curve

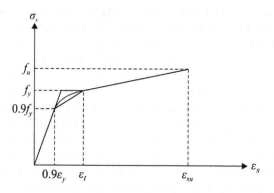

Figure 2.1 Stress–strain curves for carbon steels.

depicted in Figure 2.1 is replaced by a straight line. Liang (2009a) proposed an equation for defining the rounded part of the curve for cold-formed steel as follows:

$$\sigma_s = f_y \left(\frac{0.9\varepsilon_y - \varepsilon_s}{0.9\varepsilon_y - \varepsilon_t} \right)^{\frac{1}{45}} \quad \left(0.9\varepsilon_y < \varepsilon_s \le \varepsilon_t \right) \tag{2.1}$$

where σ_s denotes the steel stress, ε_s stands for the steel strain, f_y represents the steel yield strength, ε_y is the strain at stress f_y, and ε_t denotes the hardening strain. For mild steel, the hardening strain ε_t is taken as $10\varepsilon_y$, but for cold-formed and high-strength steels, the strain ε_t of 0.005 may be used. The ductility of cold-formed and high-strength steels is lower than that of mild steel. Therefore, the ultimate strain ε_{su} of 0.2 is specified for mild structural steel, while it is taken as 0.1 for cold-formed and high-strength steels.

The structural steel tube of a CFST column is stressed biaxially due to the concrete confinement. The hoop tension in the steel tube developed from the lateral expansion of concrete decreases the yield stress along the longitudinal direction. Therefore, the yield stress of the steel tube is reduced by the factor of 0.9 as shown in Figure 2.1.

2.3 STRESS–STRAIN RELATIONSHIPS OF STAINLESS STEELS

2.3.1 Two-stage stress–strain model by Rasmussen

Rasmussen (2003) proposed a two-stage stress–strain model for stainless steels based on the test results of tension coupons, which is illustrated

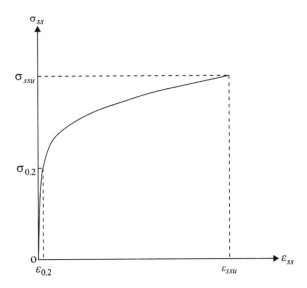

Figure 2.2 Stress–strain curve based on the two-stage model for stainless steel proposed by Rasmussen (2003).

in Figure 2.2. The two-stage model is considered in Eurocode 3 (2006) for modeling the material behavior of stainless steel. The model by Rasmussen can accurately predict the material behavior of stainless steel in tension rather than that in compression. The equation proposed by Ramberg and Osgood (1944) was adopted by Rasmussen (2003) for stainless steel stress up to the 0.2% proof stress in the first stage and is expressed by

$$\varepsilon_{ss} = \frac{\sigma_{ss}}{E_0} + 0.002\left(\frac{\sigma_{ss}}{\sigma_{0.2}}\right)^n \quad \text{for } \sigma_{ss} \leq \sigma_{0.2} \tag{2.2}$$

in which ε_{ss} represents the stainless steel strain, σ_{ss} denotes the stress at the strain ε_{ss}, E_0 stands for the Young's modulus of stainless steel, $\sigma_{0.2}$ represents the 0.2% proof stress, and n denotes the nonlinear index which is defined as

$$n = \frac{\ln(20)}{\ln(\sigma_{0.2}/\sigma_{0.01})} \tag{2.3}$$

where $\sigma_{0.01}$ denotes the 0.01% proof stress of stainless steel.

The second stage of the nonlinear stress–strain relation for stainless steel was proposed by Rasmussen (2003) as

$$\varepsilon_{ss} = 0.002 + \frac{\sigma_{0.2}}{E_0} + \varepsilon_{ssu}\left(\frac{\sigma_{ss} - \sigma_{0.2}}{\sigma_{ssu} - \sigma_{0.2}}\right)^m + \frac{\sigma_{ssu} - \sigma_{0.2}}{E_{0.2}} \qquad \text{for } \sigma_{ss} > \sigma_{0.2}$$

(2.4)

in which $E_{0.2}$ denotes the tangent modulus, which is computed by

$$E_{0.2} = \frac{E_0}{1 + 0.002nE_0/\sigma_{0.2}} \tag{2.5}$$

Rasmussen (2003) proposed an equation for determining the strain-hardening parameter m that defines the shape of stress–strain curve in the second stage. The parameter m is written as

$$m = 3.5\frac{\sigma_{0.2}}{\sigma_{ssu}} + 1 \tag{2.6}$$

In Eq. (2.4), σ_{ssu} represents the ultimate strength of stainless steel, which is determined by the following equation suggested by Rasmussen (2003):

$$\sigma_{ssu} = \left[\frac{1 - 0.0375(n - 5)}{0.2 + 185\sigma_{0.2}/E_0}\right]\sigma_{0.2} \tag{2.7}$$

The ultimate strain ε_{ssu} of stainless steel in Eq. (2.4) can be estimated by the formula derived by Rasmussen (2003) as

$$\varepsilon_{ssu} = 1 - \frac{\sigma_{0.2}}{\sigma_{ssu}} \tag{2.8}$$

2.3.2 Two-stage stress–strain model by Gardner and Nethercot

Gardner and Nethercot (2004) developed the two-stage material constitutive relation for stainless steel using the 1% proof stress rather than the ultimate strength in the second stage of the stress–strain curve. This model provides the accurate estimation of the material behavior of stainless steel up to 10% strain in tension and compression. In the first stage, Eq. (2.2) given by Ramberg and Osgood (1944) is adopted. Gardner and Nethercot (2004) derived the following expression for determining the stress–strain behavior of stainless steel in the second stage:

$$\varepsilon_{ss} = \varepsilon_{0.2} + \left(0.008 - \frac{\sigma_{1.0} - \sigma_{0.2}}{E_{0.2}}\right)\left(\frac{\sigma_{ss} - \sigma_{0.2}}{\sigma_{1.0} - \sigma_{0.2}}\right)^{n'_{0.2,1.0}} + \frac{\sigma_{ss} - \sigma_{0.2}}{E_{0.2}} \tag{2.9}$$

in which $\sigma_{1.0}$ is the 1% proof stress of stainless steel, $n'_{0.2,1.0}$ denotes the strain-hardening parameter which represents the stress–strain curve passing through the 0.2% proof stress $\sigma_{0.2}$ and 1% proof strength $\sigma_{1.0}$, and $\varepsilon_{0.2}$ represents the 0.2% proof strain which is based on the model proposed by Ramberg and Osgood (1944) and is given as

$$\varepsilon_{0.2} = \frac{\sigma_{0.2}}{E_0} + 0.002 \qquad (2.10)$$

2.3.3 Three-stage stress–strain models by Quach et al. and Abdella et al.

Experimental results show that the strain-hardening behavior of stainless steel in compression is significantly different from that in tension as illustrated in Figure 2.3 (Quach et al. 2008). This beneficial effect of significant strain-hardening of stainless steel in compression should be utilized in numerical modeling and calculations of the ultimate strengths of stainless steel members. Quach et al. (2008) proposed a three-stage stress–strain model for stainless steels in tension and compression, recognizing the difference of strain-hardening in tension and compression. An inversion of the three-stage stress–strain relationships for stainless steel given by

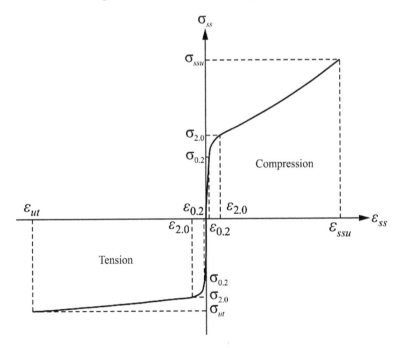

Figure 2.3 Stress–strain curve based on the three-stage model for stainless steel proposed by Quach et al. (2008).

Quach et al. (2008) was developed by Abdella et al. (2011). The inversed stress–strain relationship is easy to be implemented in numerical models for the inelastic analysis of stainless steel members. The three-stage stress–strain models for stainless steels in tension and compression are adopted in the fiber-based modeling technique discussed in this chapter.

The first stage of the three-stage stress–strain model is expressed by

$$\sigma_{ss} = \frac{E_0 \varepsilon_{ss}\left(1 + C_1 \varepsilon_r^{C_2}\right)}{1 + C_3 \varepsilon_r^{C_4} + C_1 \varepsilon_r^{C_2}} \quad \text{for } 0 \leq \varepsilon_{ss} \leq \varepsilon_{0.2} \tag{2.11}$$

where ε_r is taken as $\varepsilon_{ss}/\varepsilon_{0.2}$ and C_1, C_2, C_3, and C_4 are positive parameters. These parameters are mathematically derived by Abdella et al. (2011) as follows:

$$C_1 = \frac{\Delta}{C_2 - 1} \tag{2.12}$$

$$C_2 = 1 + \frac{B_1}{\Delta} \tag{2.13}$$

$$C_3 = G_0\left(1 + C_1\right) \tag{2.14}$$

$$C_4 = \Delta + G_1 \tag{2.15}$$

in which

$$\Delta = \frac{1 + \sqrt{1 + 4B_1}}{2} \tag{2.16}$$

$$B_1 = \frac{G_1 E_{0.2}\left(n + G_0\right)}{E_0} \tag{2.17}$$

$$G_0 = \frac{0.002 E_0}{\sigma_{0.2}} \tag{2.18}$$

$$G_1 = \frac{\varepsilon_{0.2} E_{0.2}\left(n - 1\right)}{\sigma_{0.2}} \tag{2.19}$$

Note that the elastic response of stainless steel in compression is the same as that in tension in the first stage.

In the second stage, the stress is predicted using the equations proposed by Abdella et al. (2011) as

$$\sigma_{ss} = \frac{E_{0.2}\varepsilon_* \left(1 + C_5 \varepsilon_{*r}^{C_6}\right)}{1 + C_7 \varepsilon_{*r}^{C_8} + C_5 \varepsilon_{*r}^{C_6}} + \sigma_{0.2} \quad \text{for } \varepsilon_{0.2} < \varepsilon_{ss} \le \varepsilon_{0.2} \qquad (2.20)$$

where ε_* and ε_{*r} were proposed by Abdella et al. (2011) as

$$\varepsilon_* = \varepsilon_{ss} - \varepsilon_{0.2} \qquad (2.21)$$

$$\varepsilon_{*r} = \frac{\varepsilon_{ss} - \varepsilon_{0.2}}{\varepsilon_{1.0} - \varepsilon_{0.2}} \qquad (2.22)$$

in which $\sigma_{1.0}$ is the 1.0% proof strength and $\varepsilon_{1.0}$ denotes the strain corresponding to the stress $\sigma_{1.0}$. The $\sigma_{1.0}$ stress and strain $\varepsilon_{1.0}$ in compression and tension are determined using the equations given by Quach et al. (2008) as

$$\sigma_{1.0} = \begin{cases} \left(1.085 + \dfrac{0.662}{n}\right)\sigma_{0.2} & \text{for compression} \\[2em] \left(1.072 + \dfrac{0.542}{n}\right)\sigma_{0.2} & \text{for tension} \end{cases} \qquad (2.23)$$

$$\varepsilon_{1.0} = \left[0.008 + \left(\sigma_{1.0} - \sigma_{0.2}\right)\left(\frac{1}{E_0} - \frac{1}{E_{0.2}}\right)\right] + \frac{\sigma_{1.0} - \sigma_{0.2}}{E_{0.2}} + \varepsilon_{0.2} \qquad (2.24)$$

The material parameters C_5, C_6, C_7, and C_8 in Eq. (2.20) proposed by Abdella et al. (2011) are determined by

$$C_5 = \frac{1}{C_6 - 1} \qquad (2.25)$$

$$C_6 = \frac{1}{\ln\left(\dfrac{\varepsilon_{2.0} - \varepsilon_{0.2}}{\varepsilon_{1.0} - \varepsilon_{0.2}}\right)}\left[\ln\left(1 + A_2\right) + \ln\left(\frac{H_0}{H_2}\right)\right] + C_8 \qquad (2.26)$$

$$C_7 = H_0\left(1 + C_5\right) \qquad (2.27)$$

$$C_8 = 1 + H_1 \qquad (2.28)$$

in which

$$A_2 = \frac{\left(H_2 - H_0\right)\left(n_2 - 1\right)^2}{\left(1 + n_2 H_2\right)\left(1 + n_2 H_0\right)} \qquad (2.29)$$

$$H_0 = \frac{E_{0.2}\left[(\sigma_{1.0} - \sigma_{0.2})\left(\dfrac{1}{E_0} - \dfrac{1}{E_{0.2}}\right) + 0.008\right]}{\sigma_{1.0} - \sigma_{0.2}} \tag{2.30}$$

$$H_1 = \frac{(H_0 + 1)(n_2 - 1)}{1 + n_2 H_0} \tag{2.31}$$

$$H_2 = \frac{\left(\dfrac{\varepsilon_{2.0} - \varepsilon_{0.2}}{\varepsilon_{1.0} - \varepsilon_{0.2}}\right) E_{0.2}}{\sigma_{2.0} - \sigma_{0.2}} \tag{2.32}$$

where n_2 is the material constant given by Quach et al. (2008) as follows:

$$n_2 = \begin{cases} 1.145 + 6.399\left(\dfrac{\sigma_{1.0}}{\sigma_{0.2}}\right)\left(\dfrac{E_{0.2}}{E_0}\right) & \text{for compression} \\[3mm] 1.037 + 12.255\left(\dfrac{\sigma_{1.0}}{\sigma_{0.2}}\right)\left(\dfrac{E_{0.2}}{E_0}\right) & \text{for tension} \end{cases} \tag{2.33}$$

in which $\sigma_{2.0}$ represents the 2.0% proof strength and $\varepsilon_{2.0}$ denotes the stainless steel strain at stress $\sigma_{2.0}$. The formulas for computing the stress $\sigma_{2.0}$ and strain $\varepsilon_{2.0}$ proposed by Quach et al. (2008) are expressed by

$$\sigma_{2.0} = \left[\frac{\left(\dfrac{\sigma_{1.0}}{\sigma_{0.2}} - 1\right) A^{\frac{1}{n_2}} + 1}{\dfrac{A^{\frac{1}{n_2}}}{n_2 B_2}\left(\dfrac{\sigma_{1.0}}{\sigma_{0.2}} - 1\right)\left(\dfrac{E_0}{E_{0.2}} - 1\right)\left(\dfrac{\sigma_{0.2}}{E_0}\right) + 1}\right] \sigma_{0.2} \tag{2.34}$$

$$\varepsilon_{2.0} = \left[(\sigma_{1.0} - \sigma_{0.2})\left(\frac{1}{E_0} - \frac{1}{E_{0.2}}\right) + 0.008\right]\left(\frac{\sigma_{2.0} - \sigma_{0.2}}{\sigma_{1.0} - \sigma_{0.2}}\right)^{n_2} + \varepsilon_{0.2} + \frac{\sigma_{2.0} - \sigma_{0.2}}{E_{0.2}} \tag{2.35}$$

in which

$$A = \frac{B_2}{\left(1 - \dfrac{E_0}{E_{0.2}}\right)\left(\dfrac{\sigma_{1.0}}{\sigma_{0.2}} - 1\right)\left(\dfrac{\sigma_{0.2}}{E_0}\right) + 0.008} \tag{2.36}$$

$$B_2 = \left(\frac{E_0}{E_{0.2}} - 1\right)\left(\frac{\sigma_{0.2}}{E_0}\right) + 0.018 \tag{2.37}$$

In the third stage, the stresses are calculated from strain using the following equation:

$$\sigma_{ss} = \frac{A_3 + B_3 \varepsilon_{ss}}{1 \pm \varepsilon_{ss}} \quad \text{for } \varepsilon_{ss} > \varepsilon_{2.0} \tag{2.38}$$

in which the positive and negative signs $(\pm)$ denote the tension and compression response, respectively. The constant parameters in Eq. (2.38) A_3 and B_3 are given as

$$A_3 = \sigma_{2.0}\left(1 + \varepsilon_{2.0}\right) - B_3 \varepsilon_{2.0} \tag{2.39}$$

$$B_3 = \frac{\sigma_{ssu}\left(1 + \varepsilon_{su}\right) - \sigma_{2.0}\left(1 + \varepsilon_{2.0}\right)}{\varepsilon_{ssu} - \varepsilon_{2.0}} \tag{2.40}$$

where is the ultimate strain of stainless. steel and σ_{ssu} denotes the stress at strain ε_{ssu}. Both stress σ_{ssu} and strain ε_{ssu} are computed by the equations given by Quach et al. (2008) as

$$\varepsilon_{ssu} = 1 - \frac{1}{1 + \varepsilon_{ut}} \tag{2.41}$$

$$\sigma_{ssu} = \left(1 + \varepsilon_{ut}\right)^2 \sigma_{ut} \tag{2.42}$$

where σ_{ut} represents the ultimate tensile strength and ε_{ut} denotes the strain at stress σ_{ut}, which can be computed by

$$\sigma_{ut} = \left[\frac{1 - 0.0375\left(n - 5\right)}{185\left(\frac{\sigma_{0.2}}{E_0}\right) + 0.2}\right] \sigma_{0.2} \tag{2.43}$$

$$\varepsilon_{ut} = 1 - \frac{\sigma_{0.2}}{\sigma_{ut}} \tag{2.44}$$

2.3.4 Stress–strain model by Tao and Rasmussen

Tao and Rasmussen (2016) found that the two-stage stress–strain model proposed by Rasmussen (2003) overestimated the material behavior of ferritic stainless steel. New expressions for determining the ultimate strength and strain of ferritic stainless steel were developed by Tao and Rasmussen (2016). These expressions are used in the two-stage stress–strain model given by Rasmussen (2003). The expressions for the ultimate strength of ferritic stainless steel are written as

$$\sigma_{ssu} = \begin{cases} \dfrac{\sigma_{0.2}}{0.104 + 360\,\sigma_{0.2}/E_0} & \text{for } 0.00125 \leq \sigma_{0.2}/E_0 \leq 0.00235 \\[3mm] \dfrac{\sigma_{0.2}}{0.95} & \text{for } 0.00235 < \sigma_{0.2}/E_0 \leq 0.00275 \end{cases}$$

$$(2.45)$$

The equation for the ultimate strain of ferritic stainless steel proposed by Tao and Rasmussen (2016) is

$$\varepsilon_{ssu} = 0.2 - 0.2\left(\frac{\sigma_{0.2}}{\sigma_{ssu}}\right)^{5.5} \tag{2.46}$$

2.4 STRESS–STRAIN RELATIONSHIPS OF CONCRETE

2.4.1 Compressive concrete in circular CFSST columns

The circular stainless steel tube provides confinement to the concrete core in circular CFSST columns under axial loads. The compressive strength and ductility of the concrete core in CFSST columns are found to improve. The stress–strain relationship shown in Figure 2.4 given by Liang and Fragomeni (2009) is used in the fiber model to simulate the behavior of confined concrete in circular CFSST columns. The parabolic branch OA of

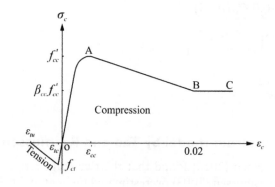

Figure 2.4 Stress–strain curve for concrete in circular CFSST columns.

the stress–strain curve is simulated using the following equation given by Mander et al. (1988) for confined concrete:

$$\sigma_c = \frac{f'_{cc}\lambda\left(\varepsilon_c/\varepsilon'_{cc}\right)}{\lambda - 1 + \left(\varepsilon_c/\varepsilon'_{cc}\right)^\lambda} \quad \text{for } 0 \le \varepsilon_c \le \varepsilon'_{cc} \tag{2.47}$$

in which σ_c stands for the compressive concrete stress in the longitudinal direction, ε_c represents the compressive concrete strain at stress σ_c, f'_{cc} and ε'_{cc} denote the compressive strength and corresponding strain of the confined concrete, and λ is determined by

$$\lambda = \frac{E_c}{E_c - \left(f'_{cc}/\varepsilon'_{cc}\right)} \tag{2.48}$$

where E_c denotes the Young's modulus of concrete which is computed using the expression suggested by ACI Committee 363 (1992) as

$$E_c = 3320\sqrt{\gamma_c f'_c} + 6900 \,(\text{MPa}) \tag{2.49}$$

in which f'_c denotes the compressive strength of concrete cylinder, γ_c represents the strength reduction parameter that considers the effects of concrete quality, loading rate, and column size on the concrete compressive strength. The parameter γ_c was proposed by Liang (2009a) as

$$\gamma_c = 1.85 D_c^{-0.135} \quad \left(0.85 \le \gamma_c \le 1.0\right) \tag{2.50}$$

where D_c stands for the concrete core diameter.

The compressive strength $\left(f'_{cc}\right)$ of confined concrete and the corresponding strain $\left(\varepsilon'_{cc}\right)$ are determined by using the equations given by Mander et al. (1988) with factor γ_c to account for the column size effect proposed by Liang and Fragomeni (2009) as follows:

$$f'_{cc} = \gamma_c f'_c + 4.1 f_{rp} \tag{2.51}$$

$$\varepsilon'_{cc} = \varepsilon'_c \left(1 + 20.5 \frac{f_{rp}}{\gamma_c f'_c}\right) \tag{2.52}$$

in which f_{rp} represents the lateral confining pressure on the concrete core and ε'_c denotes the unconfined concrete strain at stress f'_c.

Liang and Fragomeni (2009) proposed equations for computing the lateral pressure on the concrete in circular CFST columns based on the research work of Tang et al. (1996) and Hu et al. (2003). The lateral pressure on concrete depends on the diameter-to-thickness ratio (D/t), the proof stress $\sigma_{0.2}$ of stainless steel, and the concrete compressive strength f'_c. The equations proposed by Liang and Fragomeni (2009) are adopted in the fiber-based modeling technique to compute the lateral pressure on concrete in circular CFSST columns and are expressed by

$$f_{rp} = \begin{cases} 0.7(v_e - v_s)\dfrac{2t}{D-2t}\sigma_{0.2} & \text{for } \dfrac{D}{t} \leq 47 \\[4mm] \left(0.006241 - 0.0000357\dfrac{D}{t}\right)\sigma_{0.2} & \text{for } 47 < \dfrac{D}{t} \leq 150 \end{cases}$$

(2.53)

in which t is the tube thickness, D is the diameter of a circular cross-section, v_e denotes the Poisson's ratio of a stainless steel tube with infilled concrete, and v_s represents the Poisson's ratio of a hollow stainless steel tube. The Poisson's ratio v_s of the hollow stainless steel tube is taken as 0.5 at the maximum strength point. The Poisson's ratio v_e is determined using the equations suggested by Tang et al. (1996) as

$$v_e = 0.2312 + 0.3582v'_e - 0.1524\left(\frac{f'_c}{\sigma_{0.2}}\right) + 4.843v'_e\left(\frac{f'_c}{\sigma_{0.2}}\right) - 9.169\left(\frac{f'_c}{\sigma_{0.2}}\right)^2$$

(2.54)

$$v'_e = 0.881 \times 10^{-6}\left(\frac{D}{t}\right)^3 - 2.58 \times 10^{-4}\left(\frac{D}{t}\right)^2 + 1.953 \times 10^{-2}\left(\frac{D}{t}\right) + 0.4011$$

(2.55)

The values of v_e and v'_e are applicable to concrete-filled columns with $f'_c/\sigma_{0.2}$ ratios varying from 0.04 to 0.2 (Tang et al. 1996; Susantha et al. 2001).

The strain ε'_c varies from 0.002 to 0.003 with the compressive strength of concrete and can be determined by (Liang 2009a)

$$\varepsilon'_c = \begin{cases} 0.002 & \text{for } \gamma_c f'_c \leq 28 \text{ MPa} \\[4mm] 0.002 + \dfrac{\gamma_c f'_c - 28}{54000} & \text{for } 28 < \gamma_c f'_c \leq 82 \text{ MPa} \\[4mm] 0.003 & \text{for } \gamma_c f'_c > 82 \text{ MPa} \end{cases}$$

(2.56)

The linear branches AB and BC of the stress–strain curve illustrated in Figure 2.4 are simulated using the following equations:

$$\sigma_c = \begin{cases} \beta_{cc}f'_{cc} + \left(f'_{cc} - \beta_{cc}f'_{cc}\right)\left(\dfrac{\varepsilon_c - 0.02}{\varepsilon'_{cc} - 0.02}\right) & \text{for } \varepsilon'_{cc} < \varepsilon_c \leq 0.02 \\[4mm] \beta_{cc}f'_{cc} & \text{for } \varepsilon_c > 0.02 \end{cases}$$

(2.57)

where β_{cc} is the strength degradation factor for concrete, which determines the residual strength and ductility of the confined concrete in the post-peak regime. The equations proposed by Hu et al. (2003) are employed in the fiber model, expressed by

$$\beta_{cc} = \begin{cases} 1.0 & \text{for } \dfrac{D}{t} \leq 40 \\[4mm] 1.3491 + 0.0000339\left(\dfrac{D}{t}\right)^2 - 0.010085\left(\dfrac{D}{t}\right) & \text{for } 40 < \dfrac{D}{t} \leq 150 \end{cases}$$

(2.58)

2.4.2 Compressive concrete in rectangular CFSST columns

Figure 2.5 depicts the four-stage stress–strain curve suggested by Liang (2009a) for concrete in rectangular CFST columns and is adopted for concrete in rectangular CFSST columns. This constitutive model for concrete accounts for the confinement effect on the ductility of the concrete core to accurately capture the responses of rectangular CFSST columns. As illustrated in Figure 2.5, the stress–strain curve consists of a parabolic branch up to the strain ε'_c, a constant branch between the strain ε'_c and 0.005, a linear descending branch between the strains of 0.005 and 0.015, and the constant branch after the strain of 0.015. The ascending

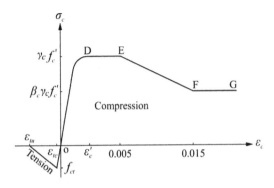

Figure 2.5 Stress–strain curve for concrete in rectangular CFSST columns.

parabolic branch OD is determined using the equation proposed by Mander et al. (1988) as

$$\sigma_c = \frac{\gamma_c f_c' \lambda (\varepsilon_c / \varepsilon_c')}{\lambda - 1 + (\varepsilon_c / \varepsilon_c')^{\lambda}} \quad \text{for } 0 \leq \varepsilon_c \leq \varepsilon_c' \tag{2.59}$$

in which λ is determined by

$$\lambda = \frac{E_c}{E_c - (\gamma_c f_c' / \varepsilon_c')} \tag{2.60}$$

The branches DE, EF, and FG of the stress–strain curve for concrete as shown in Figure 2.5 are simulated using the following equations suggested by Liang (2009a):

$$\sigma_c = \begin{cases} \gamma_c f_c' & \text{for } \varepsilon_c' < \varepsilon_c \leq 0.005 \\ \beta_c \gamma_c f_c' + 100(0.015 - \varepsilon_c)(\gamma_c f_c' - \beta_c \gamma_c f_c') & \text{for } 0.005 < \varepsilon_c \leq 0.015 \\ \beta_c \gamma_c f_c' & \text{for } \varepsilon_c > 0.015 \end{cases} \tag{2.61}$$

in which the factor β_c represents the strength degradation factor which defines the residual strength and ductility of the concrete in the post-peak range. The strength degradation factor for concrete is a function of the width-to-thickness ratio B_s / t of the rectangular column, in which B_s is the larger of B and D. Based on the test results given by Tomii and Sakino (1979), Liang (2009a) suggested that β_c can be determined as

$$\beta_c = \begin{cases} 1.0 & \text{for } \dfrac{B_s}{t} \leq 24 \\ 1.5 - \dfrac{1}{48}\dfrac{B_s}{t} & \text{for } 24 < \dfrac{B_s}{t} \leq 48 \\ 0.5 & \text{for } \dfrac{B_s}{t} > 48 \end{cases} \tag{2.62}$$

in which B and D represent the width and depth of a rectangular cross-section, respectively.

2.4.3 Concrete in tension

The concrete in tension exhibits the strain-softening and tension-stiffening behavior after cracking. This behavior is reflected by the decrease in stress after attaining the concrete tensile strength under increasing tensile strain. The tension-stiffening behavior of concrete contributes to the overall stiffness of a composite column after cracking. The material model incorporating tension-stiffening and strain-softening as depicted in Figures 2.4 and 2.5 is considered in the nonlinear analysis. The stress proportionally increases with increasing the concrete strain up to the cracking strain ε_{tc}. After attaining the strain ε_{tc}, the concrete stress decreases linearly to zero with increasing the tensile strain. The concrete tensile stress beyond the strain ε_{tu} is taken as zero. The strain ε_{tu} is determined as 10 times the cracking strain ε_{tc}. The ultimate strength f_{ct} is calculated as $0.6\sqrt{\gamma_c f_c'}$.

2.5 FIBER ELEMENT MODELING

2.5.1 Discretization of cross-sections

The numerical modeling technique presented in this chapter employs the fiber element method to discretize the cross-section of a CFSST column (Liang 2009a, b, 2011a, b). The typical fiber meshes for rectangular and circular cross-sections are shown in Figures 2.6 and 2.7, respectively. The fiber analysis incorporates the beneficial influence of the strain-hardening of stainless steel in compression and tension. In the fiber simulation, the steel–concrete composite section is first divided into fine fiber elements. The origin O of the xy coordinate system coincides with the cross-section centroid. The coordinate system is utilized to define the section geometry and eccentricity of the applied axial load. Once the cross-section has been divided into small fibers, the x and y coordinates of each fiber are determined, and the fiber areas are calculated. The contribution of each fiber is summed to obtain the axial force and moments of the cross-section under compression and bending moments.

2.5.2 Fiber strains

The stainless steel tube and concrete components of a CFSST column under axial compression is subjected to the same longitudinal strain. For columns under combined axial compression and moment, the strain is linearly varied with the depth of the column section. The strain distribution through the depth of a rectangular cross-section under axial load and biaxial bending is shown in Figure 2.6. The strain distribution within a circular cross-section is depicted in Figure 2.7. The fiber strain depends on the curvature (ϕ), the neutral axis orientation (θ), and the neutral axis depth (d_n). A circular cross-section under biaxial bending is treated as the one under uniaxial

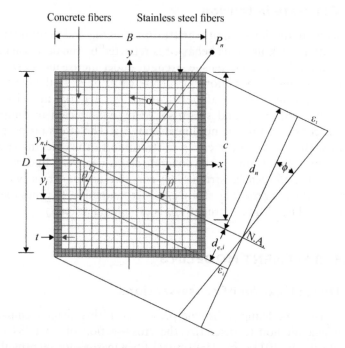

Figure 2.6 Typical fiber mesh and strain distribution in rectangular section under axial load and biaxial bending.

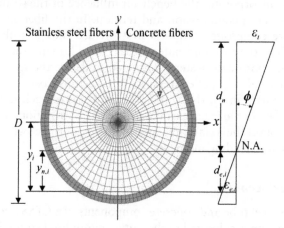

Figure 2.7 Typical fiber mesh and strain distribution in circular section under axial load and bending.

bending. In the fiber analysis, the tensile strain is assumed to be negative while the compressive strain is taken as positive. The top fiber strain ε_t depends on the neutral axis depth d_n and curvature ϕ and is computed as ϕd_n.

For $\theta = 90°$, the column cross-section is under combined concentric compression and uniaxial bending. The fiber strains are computed using the following equations derived by Liang (2009a):

$$x_{n,i} = \frac{B}{2} - d_n \tag{2.63}$$

$$d_{e,i} = \left| x_{n,i} - x_i \right| \tag{2.64}$$

$$\varepsilon_i = \begin{cases} -\phi d_{e,i} & \text{for } x_i < x_{n,i} \\ \phi d_{e,i} & \text{for } x_i \geq x_{n,i} \end{cases} \tag{2.65}$$

where $x_{n,i}$ denotes the distance from the centroid of the ith fiber element, $d_{e,i}$ denotes the perpendicular distance from the neutral axis to the fiber element centroid as shown in Figure 2.6, x_i stands for the x coordinate of the fiber i, ϕ is the curvature, and ε_i denotes the strain of fiber i.

For $0 \leq \theta < 90°$, the column cross-section is subjected to combined compression and biaxial bending. The fiber strains are computed using the formulas developed by Liang (2009a) as

$$y_{n,i} = \left(\frac{D}{2} - \frac{d_n}{\cos\theta} \right) + \tan\theta \left| \frac{B}{2} - x_i \right| \tag{2.66}$$

$$d_{e,i} = \cos\theta \left| y_{n,i} - y_i \right| \tag{2.67}$$

$$\varepsilon_i = \begin{cases} -\phi d_{e,i} & \text{for } y_i < y_{n,i} \\ \phi d_{e,i} & \text{for } y_i \geq y_{n,i} \end{cases} \tag{2.68}$$

where y_i represents the coordinate of fiber i in the y direction.

2.5.3 Axial force and bending moments

In the fiber element analysis, fiber stresses are calculated from fiber strains using the uniaxial stress–strain relationships of stainless steel and concrete presented in the preceding sections. The internal axial force and bending moments acting on the cross-section of a CFSST column are determined as stress resultants. For the cross-section of a CFSST column under combined axial compression and biaxial bending, the axial force and moments are computed by

$$P = \sum_{i=1}^{ns} \sigma_{s,i} A_{s,i} + \sum_{j=1}^{nc} \sigma_{c,j} A_{c,j} \qquad (2.69)$$

$$M_x = \sum_{i=1}^{ns} \sigma_{s,i} A_{s,i} y_i + \sum_{j=1}^{nc} \sigma_{c,j} A_{c,j} y_j \qquad (2.70)$$

$$M_y = \sum_{i=1}^{ns} \sigma_{s,i} A_{s,i} x_i + \sum_{j=1}^{nc} \sigma_{c,j} A_{c,j} x_j \qquad (2.71)$$

in which $\sigma_{s,i}$ and $\sigma_{c,j}$ represent the stresses at the stainless steel and concrete elements, respectively; $A_{s,i}$ and $A_{c,j}$ denote the stainless steel and concrete elemental areas, respectively; ns and nc stand for the numbers of stainless steel and concrete elements, respectively; M_x and M_y represent the moments about the x-axis and y-axis, respectively; x_i and y_i denote the coordinates of stainless steel fiber; x_j and y_j are the coordinates of concrete fiber.

2.5.4 Initial local buckling of stainless steel tubes

Thin stainless tube walls of rectangular CFSST columns under applied loads may undergo initial local buckling, which is characterized by deflecting outwardly from the concrete core. When a rectangular CFSST short column is subjected to concentric axial compression, the four tube walls are under uniform compressive stresses. For a rectangular CFSST short column under combined axial compression and uniaxial bending, its top flange is subjected to uniform compression while other tube walls are under either in-plane bending stresses or tension. However, when the column is subjected to combined axial load and biaxial bending, the two adjacent tube walls may be under nonuniform compressive stresses while the other adjacent walls are under in-plane bending stresses. The stress gradient coefficient (α_s) is employed to determine the stress ratio of the nonuniform compressive stresses acting on a tube wall.

The local buckling of stainless steel tube considerably reduces the strength and ductility of rectangular CFSST columns. Liang and Uy (2000) and Liang et al. (2007a) reported that the local buckling behavior of thin steel plates is affected by the initial geometric imperfection, residual stress, width-to-thickness ratio, nonuniform stress gradient, and boundary conditions of the plate. Liang et al. (2007a) proposed expressions for determining the initial local buckling stresses of steel plates in rectangular CFST beam-columns under axial compression and biaxial bending. The expressions given by Liang et al. (2007a) are adopted in the fiber element analysis to consider the influences of local buckling of stainless steel tubes on the structural responses of rectangular CFSST columns. The initial local

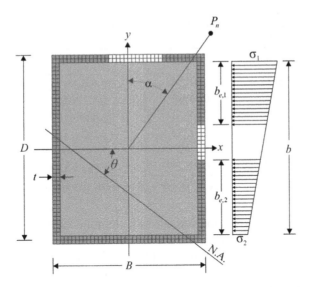

Figure 2.8 Effective widths of stainless steel tube in rectangular CFSST column section under axial load and biaxial bending.

buckling stresses of thin stainless steel plates given by Liang et al. (2007a) are determined by

$$\frac{\sigma_{1c}}{\sigma_{0.2}} = a_1 + a_2\left(\frac{b}{t}\right) + a_3\left(\frac{b}{t}\right)^2 + a_4\left(\frac{b}{t}\right)^3 \qquad \text{for } 30 \leq \frac{b}{t} \leq 100 \qquad (2.72)$$

where σ_{1c} represents the maximum stress, b stands for the clear width as illustrated in Figure 2.8, and a_1, a_2, a_3, and a_4 denote the constants. These constants depend on the stress gradient coefficient α_s, which is defined as the ratio of the minimum edge stress σ_2 to the maximum edge stress σ_1 applied to the tube wall. The constants in Eq. (2.72) are given in Table 2.1. The initial local buckling stress σ_{1c} of a tube wall under the intermediate stress gradient is estimated by linear interpolation.

2.5.5 Post-local buckling of stainless steel tubes

The behavior of thin stainless steel plates withstanding the increased load without failure after the onset of initial local buckling is called the post-local buckling (Liang 2014). Liang and Uy (2000) and Liang et al. (2007a) employed the finite element method to study the post-local buckling characteristics of carbon steel plates in CFST columns under axial compression and biaxial bending. In rectangular CFST columns, the rigid concrete core provides restraint to the steel tube walls which can only buckle locally outward. This restraint effect on the steel tube wall was simulated by the

Table 2.1 Constants for predicting the initial local buckling stress σ_{1c} of plates under stress gradients

α_s	a_1	a_2	a_3	a_4
0.0	0.6925	0.02394	-4.408×10^{-4}	1.718×10^{-6}
0.2	0.8293	0.01118	-2.427×10^{-4}	8.164×10^{-7}
0.4	0.6921	0.01223	-2.488×10^{-4}	8.676×10^{-7}
0.6	0.4028	0.02152	-3.742×10^{-4}	1.446×10^{-6}
0.8	0.5096	0.0112	-2.11×10^{-4}	7.092×10^{-7}
1.0	0.5507	0.005132	-9.869×10^{-5}	1.198×10^{-7}

Source: Adapted from Liang, Q. Q. et al., *Journal of Constructional Steel Research*, 63(3):396–405, 2007.

clamped boundary condition imposed on the four edges of a square steel tube wall and incorporating an initial geometric imperfection. A small lateral pressure was applied to the plate to induce the initial geometric imperfection. This was to ensure that the steel tube wall could only buckle outward in the nonlinear analysis. The initial geometric imperfection at the plate center was specified as $0.1t$. The compressive residual stress locked in the plate due to welding was taken as $0.25\sigma_{0.2}$. The formulas developed by Liang et al. (2007a) for computing the post-local buckling strengths of clamped steel plates are used for stainless steel tube walls in CFSST columns and are expressed by

$$\frac{\sigma_{1u}}{\sigma_{0.2}} = b_1 + b_2\left(\frac{b}{t}\right) + b_3\left(\frac{b}{t}\right)^2 + b_4\left(\frac{b}{t}\right)^3 \quad \text{for } 30 \le \frac{b}{t} \le 100 \qquad (2.73)$$

where σ_{1u} stands for the ultimate edge stress, and b_1, b_2, b_3, and b_4 are constants depending on the stress gradients and are given in Table 2.2. The ultimate stress σ_{1u} in Eq. (2.73) for the intermediate stress gradients can be calculated using the linear interpolation.

The post-local buckling strengths of thin steel plates can be determined by their effective widths (Liang 2014). Figure 2.8 illustrates the effective widths of the steel tube walls subjected to in-plane nonuniform stresses in a rectangular CFSST column. Usami (1982, 1993) proposed expressions for computing the effective width of simply supported plates in nonuniform compression. The effective width equations for simply supported plates in hollow steel tubular beam-columns under biaxial loads were derived by Shanmugam et al. (1989). Liang et al. (2007a) developed effective width formulas for clamped steel plates under nonuniform compressive stresses, which are implemented in the fiber element model to predict the effective widths of stainless steel tube walls in rectangular CFSST beam-columns. The effective widths b_{e1} and b_{e2} proposed by Liang et al. (2007a) are written as

Table 2.2 Constants for computing the post-local buckling strength of plates under stress gradients

α_s	b_1	b_2	b_3	b_4
0.0	1.257	−0.006184	1.608×10^{-4}	-1.407×10^{-6}
0.2	0.6855	0.02894	-4.89×10^{-4}	2.134×10^{-6}
0.4	0.6538	0.02888	-5.215×10^{-4}	2.424×10^{-6}
0.6	0.7468	0.01925	-3.689×10^{-4}	1.677×10^{-6}
0.8	0.6474	0.02088	-4.171×10^{-4}	2.058×10^{-6}
1.0	0.5554	0.02038	-3.944×10^{-4}	1.921×10^{-6}

Source: Adapted from Liang, Q. Q. et al., *Journal of Constructional Steel Research*, 63(3):396–405, 2007.

$$\frac{b_{e1}}{b} = \begin{cases} 0.2777 + 0.01019\left(\frac{b}{t}\right) - 1.972 \times 10^{-4}\left(\frac{b}{t}\right)^2 \\ \quad + 9.605 \times 10^{-7}\left(\frac{b}{t}\right)^3 & \text{for } \alpha_s > 0 \\ \\ 0.4186 - 0.002047\left(\frac{b}{t}\right) + 5.355 \times 10^{-5}\left(\frac{b}{t}\right)^2 \\ \quad - 4.685 \times 10^{-7}\left(\frac{b}{t}\right)^3 & \text{for } \alpha_s = 0 \end{cases}$$

(2.74)

$$\frac{b_{e2}}{b} = (2 - \alpha_s)\frac{b_{e1}}{b}$$

(2.75)

in which b_{e1} and b_{e2} represent the effective widths of the stainless steel tube wall as depicted in Figure 2.8. When the calculated effective width $(b_{e1} + b_{e2})$ is greater than the actual clear width (b) of a tube wall, the tube wall is fully effective in resisting the applied loads. For this case, the effective strength formulas given in Eq. (2.73) are employed to predict the post-local buckling strength of the stainless steel tube.

2.5.6 Modeling of progressive post-local buckling

The post-local buckling behavior of a thin plate is characterized by the redistribution of the in-plane stresses within the buckled plate (Liang 2014). The plate edges resist higher stresses while the central buckled part carries relatively lower stresses (Liang and Uy 1998). The effective width concept implies that a stainless steel plate under compressive stress gradient reaches the ultimate strength state when its maximum edge stress attains the proof stress of stainless steel. In the fiber element analysis, the fiber stresses in the effective widths are assigned to the proof stress while those in the ineffective widths are assigned to zero (Liang 2009a). After the onset of the initial local buckling, the ineffective width increases as the applied axial load increases. The ineffective width reaches its maximum value when the plate attains its ultimate strength. The maximum ineffective width of the stainless steel tube wall can be calculated as (Liang 2009a)

$$b_{ne,\max} = b - \left(b_{e1} + b_{e2}\right) \tag{2.76}$$

The ineffective width increases from zero to $b_{ne,\max}$. The intermediate value between zero and $b_{ne,\max}$ can be computed by linear interpolation based on the stress level as follows:

$$b_{ne} = \left(\frac{\sigma_1 - \sigma_{1c}}{\sigma_{0.2} - \sigma_{1c}}\right) b_{ne,\max} \tag{2.77}$$

The effective width approach is not applicable when the calculated effective width $\left(b_{e1} + b_{e2}\right)$ is greater than b. For this case, the ultimate strength formula Eq. (2.73) should be used to compute the ultimate strengths of the plate. When the maximum edge stress σ_1 exceeds the ultimate edge stress σ_{1u}, the stress σ_1 is decreased by multiplying the stress ratio σ_{1u}/σ_1 to ensure that the load carried by the plate does not exceed that computed by the ultimate strength formula (Liang 2009a).

2.6 NUMERICAL ANALYSIS PROCEDURES

2.6.1 Axial load–strain analysis

The behavior of a CFSST short column under concentric axial compression is characterized by its axial load–strain responses, which can be determined by either experiments or the nonlinear inelastic analysis on the column. The assumption of the perfect bond between the stainless steel tube and the concrete infill implies that the stainless steel tube and the concrete core are subjected to the same longitudinal axial strain. A strain-driven technique is adopted to capture the complete axial load–strain behavior of concentrically compressed short CFSST columns. In the strain-driven approach, the axial strain is incrementally increased. For a given axial strain, the fiber element stresses are calculated from the uniaxial material stress–strain relations. The axial force (P) corresponding to the given strain is computed by integrating the stresses over the cross-section. The aforementioned process is repeated until the stopping criteria are satisfied. The nonlinear analysis is terminated when the specified limiting load of $0.5P_{\max}$ or the ultimate concrete strain ε_{cu} is exceeded. The peak load on the axial load–strain curve predicted is treated as the ultimate axial load or the ultimate axial strength of the CFSST short column (Liang 2009a, 2014).

2.6.2 Moment–curvature analysis

The ultimate section moment capacity of a short CFSST beam-column under combined axial compression and bending is obtained from its complete moment–curvature curve. For a given axial load applied at a fixed angle (α) as shown in Figure 2.6, the moment–curvature curves are captured by incrementally increasing the curvature and computing the corresponding moment. For CFSST short columns subjected to biaxial loads, the neutral axis depth (d_n) and orientation (θ) are iteratively adjusted by using numerical methods such as the secant method to achieve the following equilibrium conditions (Liang 2009a, b):

$$P - P_a = 0 \tag{2.78}$$

$$\frac{M_y}{M_x} - \tan\alpha = 0 \tag{2.79}$$

where P_a stands for the applied axial load, and α represents the angle of the applied load with respect to y-axis as illustrated in Figure 2.6.

In the fiber-based analysis, the residual force at each iterative step is $r_p = P_a - P$ while the residual moment ratio is computed as $r_m = \tan\alpha - M_y/M_x$. The convergence criteria are specified as $|r_p| < \varepsilon_k$ and $|r_m| < \varepsilon_k$, where ε_k represents the convergence tolerance which is taken as 10^{-4} in the numerical analysis.

The main steps of the analysis procedure for computing the complete moment–curvature responses of a CFSST short column under axial load and biaxial bending are described as follows (Liang 2009a):

1. Input data.
2. Divide the column section into the small fibers.
3. Initialize the curvature: $\phi = \Delta\phi$.
4. Set $\theta_1 = \alpha$, $\theta_2 = \alpha/2$, $d_{n,1} = D$, and $d_{n,2} = D/2$.
5. Compute fiber strains and stresses.
6. Check local buckling and accordingly update the stainless steel fiber stresses.
7. Calculate residual force and moment ratios $r_{p,1}$, $r_{p,2}$, $r_{m,1}$, and $r_{m,2}$ corresponding to $d_{n,1}$, $d_{n,2}$, θ_1, and θ_2, respectively.
8. Compute the fiber strains and stresses.
9. Check local buckling and accordingly update steel fiber stresses.
10. Compute the axial force P.
11. Adjust the neutral axis depth d_n using the secant method.
12. Repeat Steps (8)–(11) until $|r_p| < \varepsilon_k$.

13. Compute bending moments M_x and M_y.
14. Adjust the neutral axis orientation θ using the secant algorithm.
15. Repeat Steps (8)–(14) until $|r_m| < \varepsilon_k$.
16. Calculate the resultant moment $M = \sqrt{M_x^2 + M_y^2}$.
17. Increase the curvature by $\phi = \phi + \Delta\phi$.
18. Repeat Steps (5)–(17) until $M < 0.5M_{max}$ or $\varepsilon_c > \varepsilon_{cu}$.
19. Plot the moment–curvature curve.

2.6.3 Axial load–moment interaction strength analysis

The behavior of short CFSST beam-columns under combined axial load and bending is characterized by the axial load–moment strength interaction. The presence of axial compression reduces the ultimate moment capacity of the composite section. In contrast, the presence of the bending moment reduces the ultimate axial load of the column section. The ultimate axial load (P_o) of a CFSST short column subjected to axial compression can be determined by carrying out the axial load–strain analysis on the short column. In the axial load–moment strength interaction analysis, the ultimate axial load predicted is divided into ten equal load increments. For each load increment, the ultimate moment capacity of the cross-section is computed by undertaking the moment–curvature analysis on the cross-section. The axial loads and corresponding moment capacities predicted are used to plot the axial load–moment interaction diagram of the CFSST short column.

The main steps of the analysis procedure for determining the axial load–moment strength interaction diagram of a biaxially loaded CFSST beam-column are given as follows (Liang 2009a):

1. Input data.
2. Divide the column section into the small fibers.
3. Compute the ultimate axial load P_o using the axial load–strain analysis program.
4. Set the applied load at $P_u = 0$.
5. Compute the ultimate moment capacity M_u using the moment–curvature analysis program.
6. Increase the axial load by $P_u = P_u + \Delta P_u$, in which $\Delta P_u = P_o/10$.
7. Repeat Steps (5)–(6) until $P_u > 0.9P_o$.
8. Plot the axial load–moment strength interaction diagram.

2.6.4 Solution algorithms implementing the secant method

In the nonlinear analysis of a CFSST short column under axial load and bending, the residual force and moment generated at each iteration are non-linear dynamic functions which are not derivable. The secant method is a numerical solution technique, which is used to find the roots of a nonlinear function. The main advantage of this method is that it does not require the derivative of the nonlinear function. Computational algorithms implementing the secant method have been developed by Liang (2009a, b) for determining the true orientation (θ) and depth (d_n) of the plastic neutral axis in the composite cross-section. To satisfy the force and moment equilibrium conditions, the neutral axis orientation (θ) with respect to the x-axis as illustrated in Figure 2.6 is adjusted by the following expression provided by Liang (2009a):

$$\theta_{j+2} = \theta_{j+1} - \frac{r_{m,j+1}\left(\theta_{j+1} - \theta_j\right)}{\left(r_{m,j+1} - r_{m,j}\right)} \tag{2.80}$$

in which the subscript j denotes the iteration number.

The neutral axis depth (d_n) is adjusted iteratively by using the following equation given by Liang (2009a):

$$d_{n,k+2} = d_{n,k+1} - \frac{r_{p,k+1}\left(d_{n,k+1} - d_{n,k}\right)}{\left(r_{p,k+1} - r_{p,k}\right)} \tag{2.81}$$

in which the subscript k represents the iteration number.

The convergence criterion for the orientation of the neutral axis is defined as $\left|\theta_{k+1} - \theta_k\right| \leq \varepsilon_k$, where ε_k is the convergence tolerance. The convergence criterion for the neutral axis depth is formulated as $\left|d_{n,j+1} - d_{n,j}\right| \leq \varepsilon_k$.

2.7 COMPARATIVE STUDIES

2.7.1 Validation of effective width models

The computer program developed was employed to analyze rectangular thin-walled CFSST short columns to examine the validity of the effective width models proposed by Liang et al. (2007a) for stainless steel tube walls. The numerical predictions are compared with experimental strengths and axial load–strain responses of CFSST short columns tested by Xing and Young (2018). Geometry and material parameters of tested columns are listed in Table 2.3, in which $P_{u.\exp}$ denotes the experimental ultimate axial strength, and $P_{u.\text{fib}}$ represents the predicted ultimate axial load. As appears in Table 2.3, the fiber-based simulation technique incorporating the effective width formulas for carbon steel plates developed by Liang et al. (2007a) accurately computes the ultimate axial loads of rectangular short CFSST columns. The mean ultimate axial load ratio of $P_{u.\text{fib}}/P_{u.\exp}$ is 1.02. The standard deviation and coefficient of variation are 0.081 and 0.080, respectively. It can be observed from Figure 2.9 that good agreement between the computed and experimentally measured axial load–strain responses is

Table 2.3 Ultimate axial loads of rectangular short CFSST columns under concentric compression

Specimens	$B \times D \times t$ (mm)	f'_c (MPa)	$\sigma_{0.2}$ (MPa)	E_0 (GPa)	n	$P_{u,exp}$ (kN)	$P_{u,fib}$ (kN)	$\dfrac{P_{u,fib}}{P_{u,exp}}$
SC60×40×2C30	40 × 60 × 2	31.1	599.1	198.8	5.2	370.5	367.0	0.99
SC60×40×2C30R	40 × 60 × 2	31.1	599.1	198.8	5.2	353.6	367.0	1.04
SC60×40×2C70	40 × 60 × 2	71.4	599.1	198.8	5.2	416.7	446.4	1.07
SC60×40×2C110	40 × 60 × 2	108.5	599.1	198.8	5.2	458.6	511.8	1.12
SC80×60×3C30	60 × 80 × 3	31.1	613.0	200.9	6.9	787.1	768.3	0.98
SC80×60×3C70	60 × 80 × 3	71.4	613.0	200.9	6.9	901.1	920.4	1.02
SC80×60×3C70R	60 × 80 × 3	71.4	613.0	200.9	6.9	889.9	920.4	1.03
SC80×60×3C110	60 × 80 × 3	108.5	613.0	200.9	6.9	946.3	1,060.4	1.12
SC120×60×3C30	60 × 120 × 3	31.1	610.4	205.3	6.5	914.7	779.8	0.85
SC120×60×3C70	60 × 120 × 3	71.4	610.4	205.3	6.5	1,061.4	1,011.7	0.95
Mean								1.02
Standard deviation (SD)								0.081
Coefficient of variation (COV)								0.080

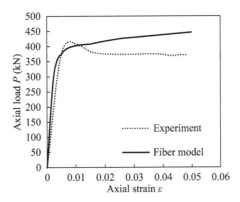

Figure 2.9 Comparison of computed and experimental axial load–strain curves for Specimen SC60×40×2C70 tested by Xing and Young (2018).

obtained. The verification indicates that the effective width models proposed by Liang et al. (2007a) for clamped carbon steel plates in the CFST columns can be incorporated in the inelastic modeling techniques to simulate local and post-local buckling of rectangular short CFSST columns.

2.7.2 Verification of the fiber element model

The experimental results provided by Lam and Gardner (2008) and Uy et al. (2011) were employed to validate the fiber-based simulation technique developed. Details on the tested CFSST short columns are given in Table 2.4. The ultimate axial loads predicted by the fiber modeling technique are compared with test results in Table 2.4. The mean computational-to-tested ultimate strength ratio is 0.97 with a standard deviation of 0.089 and a coefficient of variation of 0.091. This implies that the fiber-based model accurately determines the ultimate axial loads of concentrically compressed circular CFSST stub columns. To further verify the fiber model, the computational axial load–strain curve is compared with the experimental one in Figure 2.10. It is found that the model accurately estimates the overall trend of the load–strain relationships of the CFSST column. The measured initial axial stiffness is well captured by the numerical model. The model also simulates well the column post-yield behavior. The verification indicates that the fiber-based simulation technique accurately quantifies the overall responses of circular short CFSST columns.

2.7.3 Comparisons of stress–strain models for stainless steel

The two-stage stress–strain model proposed by Rasmussen (2003) and the three-stage stress–strain relations developed by Quach et al. (2008)

Table 2.4 Ultimate strengths of circular short CFSST columns under axial compression

Specimens	$D \times t$ (mm)	D/t	$\sigma_{0.2}$ (MPa)	E_0 (GPa)	n	f'_c (MPa)	ε_{max}	$P_{u,exp}$ (kN)	$P_{u,fib}$ (kN)	$\dfrac{P_{u,fib}}{P_{u,exp}}$
CHS 104×2-C30	104 × 2	52	412	191.9	4.3	31	0.078	699	716	1.02
CHS 104×2-C60	104 × 2	52	412	191.9	4.3	49	0.095	901	871	0.97
CHS 104×2-C100	104 × 2	52	412	191.9	4.3	65	0.044	1,133	903	0.80
CHS 114×6-C30	114.3 × 6.02	19	266	183.6	8.4	31	0.110	1,424	1,453	1.02
CHS 114×6-C60	114.3 × 6.02	19	266	183.6	8.4	49	0.163	1,648	1,859	1.13
CHS 114×6-C100	114.3 × 6.02	19	266	183.6	8.4	65	0.021	1,674	1,370	0.82
C20-50×1.2A	50.8 × 1.2	42	291	195	7	20	0.178	192	194	1.01
C20-50×1.2B	50.8 × 1.2	42	291	195	7	20	0.149	164	179	1.09
C30-50×1.2A	50.8 × 1.2	42	291	195	7	30	0.160	225	209	0.93
C20-50×1.6A	50.8 × 1.6	32	298	195	7	20	0.139	203	219	1.08
C20-50×1.6B	50.8 × 1.6	32	298	195	7	20	0.138	222	218	0.98
C30-50×1.6A	50.8 × 1.6	32	298	195	7	30	0.177	260	269	1.03
C30-50×1.6B	50.8 × 1.6	32	298	195	7	30	0.179	280	270	0.97
C20-100×1.6A	101.6 × 1.6	64	320	195	7	20	0.191	637	576	0.90
C20-100×1.6B	101.6 × 1.6	64	320	195	7	20	0.206	675	599	0.89
C30-100×1.6A	101.6 × 1.6	64	320	195	7	30	0.154	602	588	0.98
C30-100×1.6B	101.6 × 1.6	64	320	195	7	30	0.163	609	600	0.99
Mean										0.97
Standard deviation (SD)										0.089
Coefficient of variation (COV)										0.091

Source: Adapted from Patel, V.I. et al., *Journal of Constructional Steel Research*, 101:9–18, 2014.

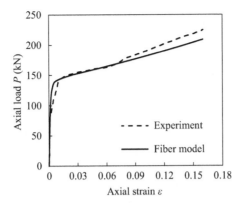

Figure 2.10 Comparison of predicted and experimental axial load-strain response of Specimen C30-50×1.2A tested by Uy et al. (2011). (Reprinted from *Journal of Constructional Steel Research*, Vol. 101, Patel, V. I., Liang, Q. Q. and Hadi, M. N. S., Nonlinear analysis of axially loaded circular concrete-filled stainless steel tubular short columns, pp. 9–18, 2014. With permission.)

and Abdella et al. (2011) for stainless steels have been implemented in the fiber-based analysis model. A circular CFSST column under axial compression was analyzed by the fiber-based model to evaluate the accuracy of the aforementioned stress–strain models for stainless steel. The column diameter was 600 mm with a diameter-to-thickness ratio of 20. The 0.2% proof stress of stainless steel tube was 320 MPa and the tube was filled with concrete with the compressive strength of 50 MPa. As demonstrated in Figure 2.11, close predictions are obtained for these stress–strain responses up to the stainless steel proof strength. This is due to the same stress–strain response of stainless steel in the elastic range. After the proof strength, a considerable difference between the axial load–strain responses predicted

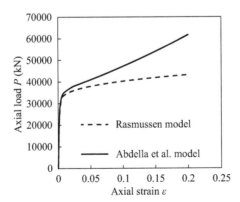

Figure 2.11 Comparisons of stress–strain models for stainless steel.

by both constitutive laws is noted. This is because stainless steel exhibits a higher strain-hardening in compression than in tension. The numerical results demonstrate that the stress–strain models for stainless steels proposed by Quach et al. (2008) and Abdella et al. (2011) should be used in the numerical modeling of CFSST columns.

2.7.4 Comparison of CFST and CFSST columns

The fiber analysis was performed to compare the structural performance of circular CFST and CFSST columns with the same dimensions. The diameter of the circular cross-section was 500 mm with a thickness of 10 mm. The yield or the proof stress of steel was 320 MPa, and compressive strength of the concrete was 50 MPa. All the parameters were kept constant in the analysis of CFST and CFSST columns. The constitutive laws for confined concrete given by Liang and Fragomeni (2009) were employed in the simulation of CFST and CFSST columns. Figures 2.12 and 2.13 show the comparison of the axial load–strain behavior and axial load–moment interaction relations for CFST and CFSST columns. As shown in Figure 2.12, CFST and CFSST beam-columns have the same initial axial stiffness. However, the axial load-carrying capacity of the CFSST column is higher than that of the CFST column. This is attributed to the higher strain-hardening of stainless steel than that of carbon steel. It can be seen from Figure 2.13 that the difference between the ultimate axial loads of the CFST column and CFSST column is larger than that between the pure bending moment capacities of the two columns. The ultimate axial strength of the CFSST column is 31.1% higher than that of the CFST column. However, the ultimate pure moment capacity of the CFSST column is only 15.8% higher than that of the CFST column.

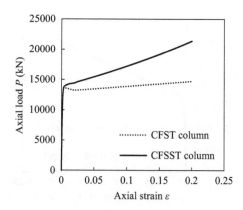

Figure 2.12 Comparison of axial load–strain curves of CFST and CFSST columns.

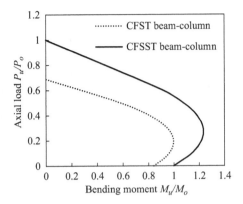

Figure 2.13 Comparison of axial load–moment interaction diagrams of CFST and CFSST beam-columns.

2.8 BEHAVIOR OF CFSST SHORT COLUMNS

The fiber-based modeling technique presented in the preceding sections was utilized to investigate the influences of material strengths and geometric parameters on the behavior of circular and rectangular short CFSST columns under axial load and bending. In the parametric studies, the effects of local buckling and concrete confinement were taken into consideration in the analyses. The Young's modulus of stainless steel was 200 GPa. The nonlinear index n of stainless steels was taken as 7.0. The ultimate concrete strain ε_{cu} was taken as 0.04 in the numerical analyses.

The stainless steel contribution ratio of a CFSST column is defined as the ratio of the ultimate axial strength of the stainless steel tube to the ultimate axial load of the composite column (Liang and Fragomeni 2009). The axial strain ductility index of a CFSST column is computed by $PI_{ad} = \varepsilon_{u.90}/\varepsilon_{y.75}$, where $\varepsilon_{y.75}$ is calculated as $\varepsilon_{0.75}/0.75$ and $\varepsilon_{0.75}$ is the strain at the axial load that attains 75% of the ultimate axial strength of the short column. The strain $\varepsilon_{u.90}$ represents the strain when the axial load falls to 90% of the ultimate axial strength of the short column (Liang and Fragomeni 2009). The curvature ductility is defined as $PI_{cd} = \phi_u/\phi_y$, in which ϕ_y is determined as $\phi_{0.75}/0.75$, where $\phi_{0.75}$ is the curvature at the moment that attains 75% of the ultimate moment capacity of the beam-column. The curvature ϕ_u denotes the curvature at the the the moment that falls to 90% of the ultimate moment capacity of the section (Liang and Fragomeni 2010).

2.8.1 Influences of depth-to-thickness ratio

The influences of D/t ratio on the performance of circular CFSST short columns were investigated by employing the fiber-based simulation method. Circular CFSST columns with a diameter of 500 mm and D/t ratios of 40,

70, and 100 were considered by varying the tube thickness. The stainless steel tubes had a proof stress of 530 MPa and were filled with concrete with a compressive strength of 50 MPa. The computed axial load–strain relations of the CFSST columns with various D/t ratios are shown in Figure 2.14. The figure illustrates that the initial axial stiffness and ultimate axial strength of the CFSST column decreases with increasing the D/t ratio. The reason for this is that increasing the D/t ratio reduces the stainless steel area and confinement effect on the concrete core. As demonstrated in Figure 2.15, increasing the D/t ratio significantly reduces the ultimate axial and bending strengths of the column. The pure moment capacity decreases slightly more than the ultimate axial strength with increasing the D/t ratio. This suggests that CFSST columns with small D/t ratios should be used to resist heavy axial loads.

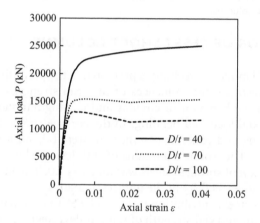

Figure 2.14 Axial load–strain responses of circular CFSST columns with various D/t ratios.

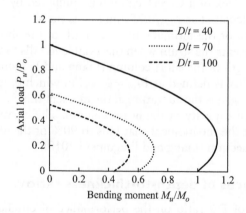

Figure 2.15 Axial load–moment interaction diagrams of CFSST beam-columns with various D/t ratios.

2.8.2 Influences of concrete strength

The influences of concrete compressive strength on the characteristics of square CFSST stub columns were studied herein. The column section was 600×600 mm and had a D/t ratio of 100. The stainless steel tube with a proof strength of 480 MPa was filled with concrete with compressive strengths of 40, 70, and 100 MPa, respectively. Figure 2.16 presents the axial load–strain curves of the short CFSST columns with various concrete strengths. The column ultimate axial load is significantly increased by increasing the concrete strength. The concrete strength has a minor effect on the initial axial stiffness of the CFSST columns. As illustrated in Figure 2.16, the higher the concrete strength, the lower the strain ductility of the column. The strain ductility indices of CFSST columns filled with 40, 70, and 100 MPa concrete are 5.133, 4.024, and 3.324, respectively. The stainless steel contribution ratio decreases from 0.439 to 0.309 and 0.238 when the concrete strength is increased from 40 MPa to 70 MPa and 100 MPa, respectively. It can be seen from Figure 2.17 that the concrete strength has the most pronounced effect on the column ultimate axial strength. The effect of concrete strength decreases with an increase in the bending moment. This implies that high-strength concrete should be used in CFSST columns to carry axial compressive loads rather than bending moments.

2.8.3 Influences of stainless steel strength

The fiber modeling program was used to examine the influences of stainless steel strength on the behavior of short circular CFSST columns. Analyses were carried out on circular columns with a diameter of 400 mm and thickness of 10 mm. The columns were made of stainless steel tubes with proof strengths of 240, 320, and 530 MPa, respectively, and filled with 80 MPa

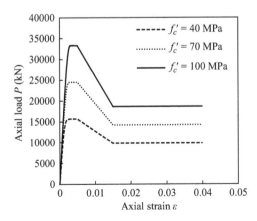

Figure 2.16 Axial load–strain responses of CFSST columns with various concrete strengths.

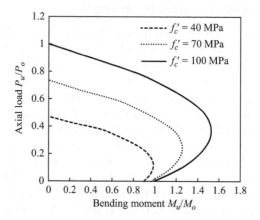

Figure 2.17 Axial load–moment interaction diagrams of short CFSST beam-columns with various concrete strengths.

concrete. The predicted load–strain responses of axially loaded CFSST short columns are given in Figure 2.18. It can be observed from Figure 2.18 that the proof stress of the stainless steel tube does not affect the column initial axial stiffness. However, the column ultimate axial load is significantly increased by using higher strength stainless steel tubes. Increasing the proof stress of the stainless steel tube from 240 to 320 and 530 MPa results in an increase in the column ultimate axial strength by 15.2% and 45.9%, respectively, and an increase in the steel contribution ratio from 0.316 to 0.355 and 0.440, respectively. The axial strain ductility of short CFSST columns with proof stresses of 240, 320, and 530 MPa are 11.01, 9.58, and 8.07, respectively. The axial load–moment interaction diagrams are provided in Figure 2.19. The ultimate axial and bending strengths of

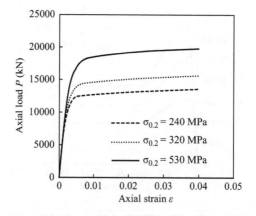

Figure 2.18 Influences of stainless steel strength on axial load–strain responses.

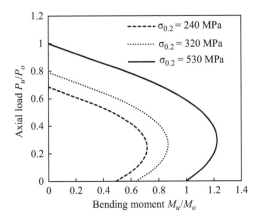

Figure 2.19 Influences of stainless steel strength on load–moment interaction diagrams.

the CFSST columns are shown to increase with increasing the proof stress. The proof stress of a stainless steel tube has the most pronounced influence on the ultimate pure moment capacity of the column. This implies that high-strength stainless steel tubes can be used to construct CFSST columns to resist large bending moments.

2.8.4 Influences of local buckling

The fiber modeling technique was used to investigate the influences of local buckling on the strength and ductility of rectangular CFSST beam-columns under axial load and bending. The geometric properties of the beam-column were: $B \times D = 400 \times 500$ mm and $D/t = 100$. The axial load was applied at angles of 0° and 45°, respectively. The rectangular tube was made of stainless steel with a proof strength of 320 MPa and filled with 65 MPa concrete. The beam-column was simulated by including or excluding local buckling effects. The influences of local buckling on the axial load–moment interaction diagram are demonstrated in Figure 2.20. The local buckling of the stainless steel tube has the pronounced influence on the column ultimate axial strength. However, its influence on the ultimate pure moment capacity is not significant. If the local buckling is ignored in the simulation, the column ultimate axial load is overestimated by 10.2%, while the pure ultimate moment capacity is overestimated by only 2.9%. As shown in Figure 2.21, for columns with a loading angle of 45°, the strengths of the column section under axial load below $0.2P_o$ are not affected by local buckling. However, at a higher loading level, local buckling reduces the column ultimate strengths. When increasing the loading angle from 0° to 45°, the maximum bending moment decreases by 15.8%.

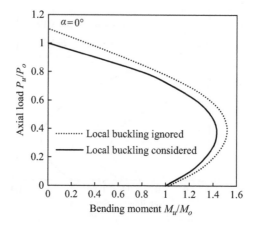

Figure 2.20 Influences of local buckling on the axial load–moment interaction diagram of CFSST short column under compression and uniaxial bending.

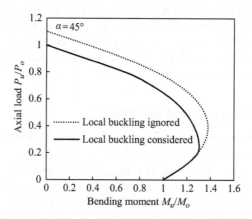

Figure 2.21 Influences of local buckling on the axial load–moment interaction diagram of CFSST short column under compression and biaxial bending.

2.8.5 Influences of section shapes

The influences of section shapes on the strength and ductility of CFSST columns with the same cross-sectional area, depth-to-thickness ratio, steel proof stress, and concrete strength were investigated. The square column had the dimensions of 500×500 mm. The diameter of the circular section was determined based on the area of square section as 564.2 mm and the D/t ratio was 50. The columns were constructed by stainless steel tubes with 320 MPa proof stress and 60 MPa concrete. The simulated axial load–strain relationships of both circular and square CFSST columns under axial loading are presented in Figure 2.22. It appears that both columns possess

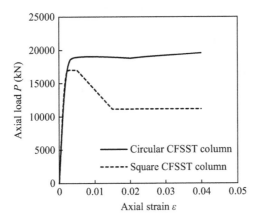

Figure 2.22 Influences of section shape on the axial load–strain behavior of CFSST short columns.

the same initial stiffness. However, the ultimate axial load of the circular section is higher than that of the square one. The square column exhibits strain-softening behavior due to the large D/t ratio of 50. In contrast, the circular column demonstrates a strain-hardening behavior. The steel contribution ratio of the circular section is 0.403, which is higher than that of the square one of 0.31. The circular section has a strain ductility of 15.85, while it is only 4.61 for the square section. This means that the circular section has better ductility than the square one.

The axial load of 10,211.67 kN, which was taken as $0.6P_o$, was applied to both circular and square beam-columns. The computed moment–curvature curves are depicted in Figure 2.23. The initial flexural stiffness

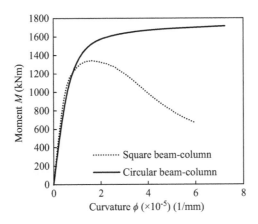

Figure 2.23 Influences of section shape on the moment–curvature responses of CFSST short columns.

of the columns is not affected by the section shape. However, the ultimate moment capacity and curvature ductility of the circular section are much higher than those of the square section. The reason for this is that the concrete confinement in a circular column increases both the column strength and ductility while the tube local buckling of a square column decreases the column strength and ductility. The curvature ductility indices of circular and square sections are 5.256 and 4.399, respectively.

2.9 DESIGN OF CFSST SHORT COLUMNS

The current design codes do not provide design methods for calculating the ultimate axial strengths of circular CFSST short columns under axial compression. Therefore, the design equations given in the current design codes for circular stub CFST columns are utilized to examine their applicability to circular CFSST short columns.

2.9.1 AISC 316-16

The design approach given in AISC 316-16 (2016) for calculating the ultimate axial strength of circular CFST columns does not consider the concrete confinement provided by the steel tubes. The ultimate axial strength is determined by adding the individual strength of steel and concrete components. In addition, AISC 316-16 limits the use of concrete with compressive strength up to 69 MPa and steel with yield strength up to 525 MPa. According to AISC 316-16, the ultimate axial strength of a circular CFSST column can be computed by

$$P_{u.\text{AISC}} = A_s\sigma_{0.2} + 0.95f_c'A_c \tag{2.82}$$

where A_s and A_c represent the areas of the stainless steel tube and concrete, respectively.

2.9.2 Eurocode 4

Eurocode 4 (2004) does not allow high-strength steel and concrete to be used to construct CFST columns. However, the concrete confinement is taken into account in the calculation of the ultimate axial strength of circular CFST columns. According to Eurocode 4, the ultimate axial strength of a circular CFSST short column under axial compression can be determined by

$$P_{u.\text{EC4}} = \eta_a A_s\sigma_{0.2} + A_cf_c'\left(1 + \eta_c\frac{t}{D}\frac{\sigma_{0.2}}{f_c'}\right) \tag{2.83}$$

The parameters η_a and η_c reflect the confinement effects and are expressed by

$$\eta_a = 0.25\left(3 + 2\overline{\lambda}\right) \quad \eta_a \leq 1 \tag{2.84}$$

$$\eta_c = 4.9 - 18.5\overline{\lambda} + 17\overline{\lambda}^2 \quad \eta_c \geq 0 \tag{2.85}$$

in which $\overline{\lambda}$ represents the relative slenderness of the composite column for the plane of bending and is determined as

$$\overline{\lambda} = \sqrt{\frac{N_{pl.Rk}}{N_{cr}}} \tag{2.86}$$

in which $N_{pl.Rk}$ stands for the cross-section plastic resistance of the CFSST column, which is expressed by

$$N_{pl.Rk} = A_s\sigma_{0.2} + A_c f_c' \tag{2.87}$$

In Eq. (2.86), N_{cr} is the Euler buckling load of the pin-ended CFSST column, written as

$$N_{cr} = \frac{\pi^2 (EI)_{eff}}{L^2} \tag{2.88}$$

in which L denotes the effective length of a CFSST column and $(EI)_{eff}$ stands for the effective flexural stiffness of the CFSST column, which is computed by

$$(EI)_{eff} = E_0 I_s + 0.6 E_{cm} I_c \tag{2.89}$$

where I_s and I_c are the second moments of area of stainless steel tube and concrete, respectively, and E_{cm} represents the modulus of elasticity for concrete, which is estimated as

$$E_{cm} = 22000\left(\frac{f_c' + 8}{10}\right)^{0.3} \tag{2.90}$$

2.9.3 Design model by Patel et al.

A design formula was proposed Liang and Fragomeni (2009) for the design of axially loaded circular CFST short columns including concrete confinement effects. Their design equation is applicable to CFST columns made of concrete with compressive strength up to 120 MPa and steel tubes with yield stress up to 690 MPa. The design model proposed by Liang and Fragomeni (2009) was modified by Patel et al. (2014) for the design of circular CFSST short columns under axial compression as follows:

$$P_{u.\text{Patel}} = \left(\gamma_c f_c' + 4.1 f_{rp} \right) A_c + \gamma_s \sigma_{0.2} A_s \tag{2.91}$$

in which γ_c is determined using Eq. (2.50), f_{rp} is calculated using Eq. (2.53) and γ_s represents the strength factor for the tube, which was proposed by Patel et al. (2014) for stainless steel tubes as

$$\gamma_s = \begin{cases} 3\left(\dfrac{D}{t}\right)^{-0.1} & \text{for } \dfrac{D}{t} < 80 \\[4mm] 2\left(\dfrac{D}{t}\right)^{-0.1} & \text{for } \dfrac{D}{t} \geq 80 \end{cases} \tag{2.92}$$

The experimental axial loads of the tested CFSST short columns are compared with those calculated using the above-mentioned design methods given in current design codes in Table 2.5. The mean ratios of the ultimate axial loads calculated by the methods given in AISC 316-16 (2016) and Eurocode 4 (2004) to the tested values are 0.55 and 0.76, respectively. The design methods for CFST columns provided in AISC 316-16 and Eurocode 4 significantly underestimate the ultimate sterngths of ciruclar CFSST short columns. This is because Eurocode 4 does not consider the stainless steel strain-hardening of stainless steel and AISC 316-16 does not account for both strain-hardening and concrete confinement. It can be seen that the design model proposed by Patel et al. (2014) including the effects of strain-hardening and concrete confinement accurately determines the ultimate axial strengths of circular CFSST short columns.

2.10 CONCLUSIONS

The fiber-based modeling technique has been presented in this chapter that simulates the behavior of short CFSST columns under axial and eccentric loads. The modeling technique incorporates not only the accurate stress–strain models for stainless steel and confined concrete but also the effects of local buckling, concrete confinement, and high-strength materials. Computational algorithms implementing the secant method are employed to solve the nonlinear dynamic equilibrium equations of CFSST beam-columns subjected to eccentric loading. The computational procedures for accurately capturing the axial load–strain responses, moment–curvature relations, and axial load–moment strength interaction diagrams for short CFSST columns have been described.

Table 2.5 Comparison of experimental ultimate axial strengths of circular short CFSST columns with those calculated by design models

Specimens	$P_{u.exp}$ (kN)	AISC 316-16		Eurocode 4		Patel et al. (2014)	
		$P_{u.AISC}$ (kN)	$\dfrac{P_{u.AISC}}{P_{u.exp}}$	$P_{u.EC4}$ (kN)	$\dfrac{P_{u.EC4}}{P_{u.exp}}$	$P_{u.Patel}$	$\dfrac{P_{u.Patel}}{P_{u.exp}}$
CHS 104×2-C30	699	495	0.71	683	0.99	834	1.19
CHS 104×2-C60	901	630	0.70	818	0.91	974	1.08
CHS 104×2-C100	1,133	749	0.66	939	0.83	1,099	0.97
CHS 114×6-C30	1,424	787	0.55	1,138	0.80	1,632	1.02
CHS 114×6-C60	1,648	927	0.56	1,279	0.77	1,800	1.09
CHS 114×6-C100	1,674	1,052	0.63	1,405	0.84	1,898	1.13
C20-50×1.2A	192	89	0.47	129	0.67	173	0.90
C20-50×1.2B	164	89	0.54	129	0.78	173	1.06
C30-50×1.2A	225	107	0.47	146	0.65	199	0.89
C20-50×1.6A	203	108	0.53	158	0.78	220	1.08
C20-50×1.6B	222	108	0.48	158	0.71	220	0.99
C30-50×1.6A	260	124	0.48	175	0.67	247	0.95
C30-50×1.6B	280	124	0.44	175	0.62	247	0.88
C20-100×1.6A	637	305	0.48	426	0.67	532	0.84
C20-100×1.6B	675	305	0.45	426	0.63	532	0.79
C30-100×1.6A	602	378	0.63	499	0.83	608	1.01
C30-100×1.6B	609	378	0.62	499	0.82	608	1.00
Mean			0.55		0.76		0.99
Standard deviation (SD)			0.089		0.103		0.108
Coefficient of variation (COV)			0.162		0.136		0.108

The following conclusions are deduced from the comparative and parametric studies:

- The design formulas proposed by Liang et al. (2007a) for carbon steel plates can be used to determine the initial local and post-local buckling strengths of stainless steel tube walls in rectangular CFSST columns with good accuracy.
- The fiber-based analysis technique presented accurately simulates the behavior of CFSST short columns.
- The three-stage stress–strain models proposed by Quach et al. (2008) and Abdella et al. (2011) recognize the different strain-hardening characteristics of stainless steel in tension and compression so that they are recommended to be used in the nonlinear analysis of CFSST columns.

- For columns with the same dimensions, concrete strength, and steel yield strength, CFSST columns have higher strength and ductility than CFST columns.
- Increasing the D/t ratio reduces the strength and ductility of CFSST columns.
- The concrete strength has the most pronounced effect on the ultimate axial loads of CFSST columns, and its effect decreases with increasing the bending moment.
- The proof stress of stainless steel has the most pronounced effect on the ultimate pure moment of CFSST columns, and its effect is decreased by increasing the axial load.
- The nonlinear analysis procedures may overestimate the ultimate strengths of CFSST short columns if local buckling is not considered.
- The overall performance of circular CFSST columns is better than that of rectangular CFSST columns.
- The design methods given in AISC 316-16 (2016) and Eurocode 4 (2004) significantly underestimate the axial load-carrying capacities of circular CFSST short columns. The design model proposed by Patel et al. (2014) yields accurate calculations of the ultimate axial strengths of circular CFSST columns.

REFERENCES

Abdella, K., Thannon, R. A., Mehri, A. I. and Alshaikh, F. A. (2011) "Inversion of three-stage stress-strain relation for stainless steel in tension and compression," *Journal of Constructional Steel Research*, 67(5): 826–832.

ACI Committee 363 (1992) "State of the art report on high-strength concrete," ACI Publication 363R-92. Detroit, MI, USA: American Concrete Institute.

AISC 316-16 (2016) *Specification for Structural Steel Buildings*, Chicago, IL, USA: AISC.

Bridge, R. Q. and O'Shea, M. D. (1998) "Behaviour of thin-walled steel box sections with or without internal restraint," *Journal of Constructional Steel Research*, 47(1–2): 73–91.

Ellobody, E. and Young, B. (2006) "Design and behaviour of concrete-filled cold-formed stainless steel tube columns," *Engineering Structures*, 28(5): 716–728.

Ellobody, E., Young, B. and Lam, D. (2006) "Behaviour of normal and high strength concrete-filled compact steel tube circular stub columns," *Journal of Constructional Steel Research*, 62(7): 706–715.

Eurocode 3 (2006) *Design of Steel Structures – Part 1-4: General Rules – Supplementary Rules for Stainless Steels*, Brussels, Belgium: European Committee for Standardization, CEN.

Eurocode 4 (2004) *Design of Composite Steel and Concrete Structures, Part 1-1: General Rules and Rules for Buildings*, Brussels, Belgium: European Committee for Standardization, CEN.

Furlong, R. W. (1967) "Strength of steel-encased concrete beam columns," *Journal of the Structural Division*, ASCE, 93(5): 113–124.

Gardner, L. and Nethercot, D. A. (2004) "Numerical modeling of stainless steel structural components – A consistent approach," *Journal of Structural Engineering*, ASCE, 130(10): 1586–1601.

Ge, H. B. and Usami, T. (1992) "Strength of concrete-filled thin-walled steel box columns: Experiments," *Journal of Structural Engineering*, ASCE, 118(11): 3036–3054.

Giakoumelis, G. and Lam, D. (2004) "Axial capacity of circular concrete-filled tube columns," *Journal of Constructional Steel Research*, 60(7): 1049–1068.

Han, L. H. (2002) "Tests on stub columns of concrete-filled RHS sections," *Journal of Constructional Steel Research*, 58(3): 353–372.

Hassanein, M. F., Kharoob, O. F. and Liang, Q. Q. (2013) "Behaviour of circular concrete-filled lean duplex stainless steel tubular short columns," *Thin-Walled Structures*, 68: 113–123.

Hu, H. T., Huang, C. S., Wu, M. H. and Wu, Y. M. (2003) "Nonlinear analysis of axially loaded concrete-filled tube columns with confinement effect," *Journal of Structural Engineering*, ASCE, 129(10): 1322–1329.

Klöppel, V. K. and Goder, W. (1957) "An investigation of the load carrying capacity of concrete filled steel tubes and development of design formula," *Der Stahlbau*, 26(2): 44–50.

Knowles, R. B. and Park, R. (1969) "Strength of concrete-filed steel tubular columns," *Journal of the Structural Division*, ASCE, 95(12): 2565–2587.

Lam, D. and Gardner, L. (2008) "Structural design of stainless steel concrete filled columns," *Journal of Constructional Steel Research*, 64(11): 1275–1282.

Liang, Q. Q. (2009a) "Performance-based analysis of concrete-filled steel tubular beam-columns, Part I: Theory and algorithms," *Journal of Constructional Steel Research*, 65(2): 363–372.

Liang, Q. Q. (2009b) "Performance-based analysis of concrete-filled steel tubular beam-columns, Part II: Verification and applications," *Journal of Constructional Steel Research*, 65(2): 351–362.

Liang, Q. Q. (2009c) "Strength and ductility of high strength concrete-filled steel tubular beam-columns" *Journal of Constructional Steel Research*, 65(3): 687–698.

Liang, Q. Q. (2011a) "High strength circular concrete-filled steel tubular slender beam-columns, Part I: Numerical analysis," *Journal of Constructional Steel Research*, 67(2): 164–171.

Liang, Q. Q. (2011b) "High strength circular concrete-filled steel tubular slender beam-columns, Part II: Fundamental behavior," *Journal of Constructional Steel Research*, 67(2): 172–180.

Liang, Q. Q. (2014) *Analysis and Design of Steel and Composite Structures*, Boca Raton, FL, USA and London, UK: CRC Press, Taylor & Francis Group.

Liang, Q. Q. and Fragomeni, S. (2009) "Nonlinear analysis of circular concrete-filled steel tubular short columns under axial loading," *Journal of Constructional Steel Research*, 65(12): 2186–2196.

Liang, Q. Q. and Fragomeni, S. (2010) "Nonlinear analysis of circular concrete-filled steel tubular short columns under eccentric loading," *Journal of Constructional Steel Research*, 66(2): 159–169.

Liang, Q. Q. and Uy, B. (1998) "Parametric study on the structural behaviour of steel plates in concrete-filled fabricated thin-walled box columns," *Advances in Structural Engineering*, 2(1): 57–71.

Liang, Q. Q. and Uy, B. (2000) "Theoretical study on the post-local buckling of steel plates in concrete-filled box columns," *Computers and Structures*, 75(5): 479–490.

Liang, Q. Q., Uy, B. and Liew, J. Y. R. (2006) "Nonlinear analysis of concrete-filled thin-walled steel box columns with local buckling effects," *Journal of Constructional Steel Research*, 62(6): 581–591.

Liang, Q. Q., Uy, B. and Liew, J. Y. R. (2007a) "Local buckling of steel plates in concrete-filled thin-walled steel tubular beam-columns," *Journal of Constructional Steel Research*, 63(3): 396–405.

Liang, Q. Q., Uy, B. and Liew, J. Y. R. (2007b) "Strength of concrete-filled steel box columns with buckling effects," *Australian Journal of Structural Engineering*, 7(2): 145–155.

Liang, Q. Q., Uy, B., Wright, H. D. and Bradford, M. A. (2004) "Local buckling of steel plates in double skin composite panels under biaxial compression and shear," *Journal of Structural Engineering*, ASCE, 130(3): 443–451.

Mander, J. B., Priestly, M. N. J. and Park, R. (1988) "Theoretical stress-strain model for confined concrete," *Journal of Structural Engineering*, ASCE, 114(8): 1804–1826.

O'Shea, M. D. and Bridge, R. Q. (2000) "Design of circular thin-walled concrete filled steel tubes," *Journal of Structural Engineering*, ASCE, 126(11): 1295–1303.

Patel, V. I., Liang, Q. Q. and Hadi, M. N. S. (2014) "Nonlinear analysis of axially loaded circular concrete-filled stainless steel tubular short columns," *Journal of Constructional Steel Research*, 101: 9–18.

Quach, W. M., Teng, J. G. and Chung, K. F. (2008) "Three-stage full-range stress-strain model for stainless steels," *Journal of Structural Engineering*, ASCE, 134(9): 1518–1527.

Ramberg, W. and Osgood, W. R. (1944) "Description of stress-strain relations from offset yield strength values," NACA Technical Note no. 927.

Rasmussen, K. J. R. (2003) "Full-range stress-strain curves for stainless steel alloys," *Journal of Constructional Steel Research*, 59(1): 47–61.

Sakino, K., Nakahara, H., Morino, S. and Nishiyama, I. (2004) "Behavior of centrally loaded concrete-filled steel-tube short columns," *Journal of Structural Engineering*, ASCE, 130(2): 180–188.

Schneider, S. P. (1998) "Axially loaded concrete-filled steel tubes," *Journal of Structural Engineering*, ASCE, 124(10): 1125–1138.

Shanmugam, N. S., Liew, J. Y. R. and Lee, S. L. (1989) "Thin-walled steel box columns under biaxial loading," *Journal of Structural Engineering*, ASCE, 115(11): 2706–2726.

Susantha, K. A. S., Ge, H. B. and Usami, T. (2001) "Uniaxial stress-strain relationship of concrete confined by various shaped steel tubes," *Engineering Structures*, 23(10): 1331–1347.

Tang, J., Hino, S., Kuroda, I. and Ohta, T. (1996) "Modeling of stress-strain relationships for steel and concrete in concrete filled circular steel tubular columns," *Steel Construction Engineering*, JSSC, 3(11): 35–46.

Tao, Z. and Han, L. H. (2006) "Behaviour of concrete-filled double skin rectangular steel tubular beam-columns," *Journal of Constructional Steel Research*, 62(7): 631–646.

Tao, Z. and Rasmussen, K. J. R. (2016) "Stress-strain model for ferritic stainless steels," *Journal of Materials in Civil Engineering*, ASCE, 28(2): 06015009.

Tao, Z., Uy, B., Liao, F. Y. and Han, L. H. (2011) "Nonlinear analysis of concrete-filled square stainless steel stub columns under axial compression," *Journal of Constructional Steel Research*, 67(11): 1719–1732.

Tao, Z., Wang, Z. B. and Yu, Q. (2013) "Finite element modelling of concrete-filled steel stub columns under axial compression," *Journal of Constructional Steel Research*, 89: 121–131.

Thai, H. T., Uy, B., Khan, M., Tao, Z. and Mashiri, F. (2014) "Numerical modelling of concrete-filled steel box columns incorporating high strength materials," *Journal of Constructional Steel Research*, 102: 256–265.

Tomii, M. and Sakino, K. (1979) "Elastic-plastic behavior of concrete filled square steel tubular beam-columns," *Transactions of the Architectural Institute of Japan*, 280: 111–120.

Usami, T. (1982) "Post-buckling of plates in compression and bending," *Journal of the Structural Division*, ASCE, 108(3): 591–609.

Usami, T. (1993) "Effective width of locally buckled plates in compression and bending," *Journal of Structural Engineering*, ASCE, 119(5): 1358–1373.

Uy, B. (1998) "Local and post-local buckling of concrete filled steel welded box columns," *Journal of Constructional Steel Research*, 47(1–2): 47–72.

Uy, B. (2000) "Strength of concrete-filled steel box columns incorporating local buckling," *Journal of Structural Engineering*, ASCE, 126(3): 341–352.

Uy, B. (2001) "Strength of short concrete filled high strength steel box columns," *Journal of Constructional Steel Research*, 57(2): 113–134.

Uy, B., Tao, Z. and Han, L. H. (2011) "Behaviour of short and slender concrete-filled stainless steel tubular columns," *Journal of Constructional Steel Research*, 67(3): 360–378.

Wright, H. D. (1995) "Local stability of filled and encased steel sections," *Journal of Structural Engineering*, ASCE, 121(10): 1382–1388.

Xing, B. and Young, B. (2018) "Experimental investigation of concrete-filled lean duplex stainless steel RHS stub columns," *Tubular Structures*, XVI: 95–99.

Young, B. and Ellobody, E. (2006) "Experimental investigation of concrete-filled cold-formed high strength stainless steel tube columns," *Journal of Constructional Steel Research*, 62(5): 484–492.

Chapter 3

Nonlinear analysis of circular CFSST slender columns

3.1 INTRODUCTION

Circular concrete-filled stainless steel tubular (CFSST) slender columns are increasingly utilized in commercial and office composite structures to support vertical loads as well as lateral loads arising from wind and earthquakes (Ellobody 2013a; Han et al. in press). When the column slenderness ratio (L/r) is greater than 22, the column is classified as a slender column. As discussed in Chapter 2, the ultimate strength of a short CFSST column is governed by the geometric and material properties of its cross-section. However, the ultimate strength of a CFSST slender column is generally governed by its overall stability. The increase in the length of a composite column reduces its stability and ultimate strength. Therefore, the nonlinear analysis and design of slender columns must take the stability effects into account. Practical columns are generally subjected to combined actions of axial load and bending which are induced by the frame actions. No experimental studies on the behavior of circular slender CFSST beam-columns under eccentric loading have been reported. The current design codes, such as Eurocode 4 (2004), AISC 360-16 (2016), and AS/NZS 5100.6 (2017), do not provide design guidelines for slender CFSST columns. Therefore, numerical investigations into the behavior of circular slender CFSST columns under eccentric loading are much needed.

Experiments on the ultimate strengths of circular slender concrete-filled steel tubular (CFST) columns made of carbon steel tube were undertaken by Neogi et al. (1969), Knowles and Park (1969), Rangan and Joyce (1992), and Portolés et al. (2011a). However, experimental investigations on the fundamental behavior of circular slender CFSST columns have been very limited. After reviewing the recent research on the performance of short and slender CFSST columns, Han et al. (in press) reported that the design methods given in current design standards for CFST columns made of carbon steel tubes significantly underestimate the ultimate axial strengths of CFSST columns. This is because these codified design methods do not consider the strain-hardening of stainless steel in tension and compression. Tests on pin-ended circular slender CFSST columns under axial compression were conducted by Uy et al. (2011). It was observed that all slender CFSST columns failed by the overall column buckling with large lateral deflections. In addition, it was reported that the current design codes yield conservative ultimate strength predictions of concentrically loaded circular slender CFSST columns.

Numerical analyses on circular slender CFST columns were undertaken by Shanmugam et al. (2002), Hatzigeorgiou (2008), Liang (2011a, b), and Portolés et al. (2011b) and on concrete-encased composite columns by Ky et al. (2015). Ellobody (2013b) developed a finite element model using the general purpose finite element software Abaqus for predicting the behavior of circular slender CFSST columns under eccentric loading. Both the measured stress–strain responses and the two-stage stress–strain model proposed by Rasmussen (2003) were employed to model stainless steel. The same strain-hardening of stainless steel in tension and compression was assumed in the stress–strain relationships used in the finite element modeling, which underestimated the ultimate axial strengths of slender CFSST columns. A fiber-based numerical model was developed by Patel et al. (2017) for computing the axial load–deflection and axial load–moment interaction responses of eccentrically loaded circular slender CFSST beam-columns. The model accounted for the different strain-hardening characteristics of stainless steel in tension and compression as determined by the three-stage stress–strain models given by Quach et al. (2008) and Abdella et al. (2011). The developed fiber-based analysis technique predicted well the experimentally measured responses of CFSST slender columns.

This chapter presents the numerical analysis, behavior, and design of circular slender CFSST columns subjected to combined axial load and bending. The mathematical model for the nonlinear analysis of CFSST slender columns utilizes the fiber approach to discretize the cross-section. The accurate material constitutive models proposed by Quach et al. (2008) and Abdella et al. (2011) for stainless steel with different strain-hardening responses in tension and compression are incorporated in the mathematical model. The influences of concrete confinement, second order, out-of-straightness, and inelastic material behavior are taken into consideration. The formulation of the mathematical model as well as computational procedures for axial load–deflection curves and axial load–moment interaction diagrams of CFSST slender columns are provided. The mathematical model verified by test data is employed to investigate the behavior of circular slender CFSST columns with various geometric and material parameters. The design of circular slender CFSST columns under axial compression in accordance with Eurocode 4 (2004) is discussed.

3.2 MODELING OF CROSS-SECTIONS

The cross-section behavior in terms of moment–curvature relations is a prerequisite for the stability simulation of eccentrically loaded circular slender

CFSST beam-columns. As discussed in Chapter 2, the cross-section of a circular CFST column is modeled by discretizing the cross-section into fine fiber elements, allowing for different materials to be included in the cross-section as illustrated in Figure 2.7. The material constitutive relations of stainless steel and concrete given in Chapter 2 are used in the simulation of the cross-sections of slender columns. The material models account accurately for the elastic–plastic responses of stainless steel with different strain-hardening behaviors in tension and compression, the nonlinear responses of concrete in compression, and concrete tensile cracking. In the moment–curvature simulation, the neutral axis depth within the cross-section is initialized. For a given curvature ϕ, the strain ε_t at the extreme compression fiber is calculated using the expression of $\varepsilon_t = \phi d_n$. The fiber stresses in concrete and stainless steel are computed from their strains by means of applying the uniaxial stress–strain relationships. The equilibrium condition between the internal axial force and the applied axial load is established by means of iteratively adjusting the neutral axis depth d_n using either the secant method (Liang 2011a, b, 2014) or Müller's numerical technique (Patel et al. 2012a, b). The internal section moment is calculated by the numerical integration of stainless steel and concrete stresses over the cross-section. The above computational scheme is repeated until the complete axial load–moment–curvature relationship is established.

3.3 MODELING OF LOAD–DEFLECTION RESPONSES

3.3.1 Mathematical formulation

The mid-height deflection of a CFSST slender column under applied load can be computed by using the load control method. However, the load control method cannot predict the post-peak branch of the axial load–deflection curve. Therefore, the deflection control method, in which the load is computed for a given mid-height deflection, is implemented to determine the complete load–deflection curve including the post-peak responses. The current design standards use pin-ended supports to design slender beam-columns. The pin-ended beam-column shown in Figure 3.1 is treated in the formulation of the slender column model. The beam-column subjected to the equal eccentric loads at both ends bends into single curvature. In Figure 3.1, e represents the loading eccentricity, L denotes the column length, and u_m stands for the mid-height deflection. The out-of-straightness as initial imperfection u_o in the form of deflected shape shown in Figure 3.1 is taken into account in the beam-column model. The second-order effects, which are caused by the interaction of the axial load and lateral deflections, amplify the bending moment and are included in the column stability analysis.

In order to compute the deflections, the deflected shape of slender CFSST beam-columns under applied loads needs to be determined. The part-cosine wave shape function was used to model the deflected shape of

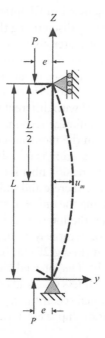

Figure 3.1 The pin-ended beam-column model. (Reprinted from *Journal of Constructional Steel Research*, Vol. 130, Patel, V. I., Liang, Q. Q. and Hadi, M. N. S., Nonlinear analysis of circular high strength concrete-filled stainless steel tubular slender beam-columns, pp. 1–13, 2017. With permission.)

circular slender CFST beam-columns by Neogi et al. (1969), who reported that this shape function sufficiently estimated the deflected shape. Other shape functions have been employed by researchers, including the part-sine wave function suggested by Shakir-Khalil and Zeghiche (1989), the fourth-degree function utilized by El-Tawil et al. (1995), and the second-degree parabola function implemented by Diniz and Frangopol (1997). The present fiber-based mathematical model uses the part-sine shape function to determine the deflections of pin-ended circular slender CFSST beam-columns as illustrated in Figure 3.1. The deflection u along the column length at any point (y, z) can be written as

$$u = u_m \sin \frac{\pi z}{L} \tag{3.1}$$

The initial geometric imperfection (u_{oy}) of the column is also defined by the displacement equation as

$$u_{oy} = u_o \sin \frac{\pi z}{L} \tag{3.2}$$

where u_o represents the initial geometric imperfection presented at the column mid-height.

The internal section moment is calculated by means of the moment–curvature modeling scheme. To determine internal moment, an expression to predict the mid-height curvature as a function of deflection is needed. This expression can mathematically be derived by performing the double differentiation of Eq. (3.1) as

$$\phi = \frac{\partial^2 u}{\partial z^2} = \frac{\pi^2}{L^2} u_m \sin \frac{\pi z}{L} \tag{3.3}$$

The curvature at the column mid-height as a function of mid-height deflection and column length is determined from Eq. (3.3) as

$$\phi_m = u_m \frac{\pi^2}{L^2} \tag{3.4}$$

The formulation accounts for the effects of the loading eccentricity (e), initial member imperfection (u_o), and second order on the nonlinear responses of slender columns under eccentric loading. The interaction of axial load and mid-height deflection (u_m) amplifies the bending moment on the column length, which is the second-order effect. The external bending moment at the column mid-height is therefore

$$M_{me} = P(u_o + u_m + e) \tag{3.5}$$

In the modeling of axial load–deflection responses, for a given deflection increment at the column mid-height, the internal axial force and moment are computed, and these generalized section forces must satisfy the force and moment equilibrium conditions at the column mid-height. This implies that the internal force that satisfies the moment equilibrium at the column mid-height is determined as the applied axial load at an eccentricity of e at the column ends. To achieve the equilibrium condition in the incremental and iterative numerical analysis, the residual moment expressed by the following equation must be sufficiently small:

$$r_{pu} = M_{mi} - P(u_o + u_m + e) \tag{3.6}$$

Equation (3.6) is a dynamic function which changes as the variables change in the iterative modeling process. When $|r_{pu}| < \varepsilon_k$, the moment equilibrium at the column mid-height is achieved. The convergence tolerance ε_k is taken as 10^{-4}. Computational algorithms are developed by means of implementing Müller's numerical approach to solve the above dynamic function (Müller 1956).

3.3.2 Computational procedure

An iterative computational procedure has been developed that calculates the load–deflection responses of eccentrically loaded circular slender CFSST beam-columns. The computational procedure accounts for the influences of concrete confinement, second order, and initial geometric imperfection as discussed in the preceding section. The computation starts with initializing the deflection (u_m) at the column mid-height. For a given deflection increment, the curvature at the column mid-height is computed. The strains of fiber elements in the cross-section are then determined. The stresses of fiber elements are therefore computed from fiber strains. The internal axial force and moment are computed. The above process is repeated until the moment equilibrium condition is achieved at the column mid-height. The deflection is increased and the modeling process is repeated until the complete axial load–deflection curve of the slender CFSST column is obtained (Liang 2011a, b).

The computational procedure proposed by Liang (2011a) is described as follows:

1. Input data.
2. Mesh the column cross-section with fine fiber elements and calculate their coordinates and areas.
3. Initialize the mid-height deflection as $u_m = \Delta u_m$.
4. Determine the curvature ϕ_m for the given deflection u_m.
5. Assign three initial values to the neutral axis depths as $d_{n,1} = D/4$, $d_{n,2} = D/2$, and $d_{n,3} = D$.
6. Calculate element stresses from their strains by means of material constitutive models for stainless steel and concrete.
7. Compute the internal axial force P and moment M_{mi} by means of integrating stresses over the cross-section for each neutral axis depth.
8. Determine three residual moments $r_{pu,1}$, $r_{pu,2}$, and $r_{pu,3}$ that correspond to three neutral axis depths, respectively.
9. Adjust the neutral axis depth d_n by using the numerical solution algorithms.
10. Compute element stresses from their strains using material constitutive models for stainless steel and concrete.
11. Determine the internal axial force P and moment M_{mi} by integrating stresses over the cross-section.

12. Calculate the residual moment r_{pu} for the neutral axis depth d_n.
13. Repeat Steps (9)–(12) until the equilibrium condition is satisfied.
14. Increase the mid-height deflection u_m by using $u_m = u_m + \Delta u_m$.
15. Repeat Steps (4)–(14) until the computed axial load P is less than $0.65 P_{max}$ or the mid-height deflection limit is exceeded, where the maximum axial load P_{max} is taken as the column ultimate axial load.
16. Plot the axial load–deflection curve.

3.4 GENERATING AXIAL LOAD–MOMENT STRENGTH ENVELOPES

3.4.1 Mathematical modeling

The computational procedure for axial load–deflection responses can be employed to generate the axial load–moment strength interaction diagrams for slender CFSST beam-columns under eccentric loading by means of changing the eccentricity of the applied load. The axial load–moment strength interaction diagrams are strength envelopes which are used as yield surfaces in the advanced analysis of composite frames. The ultimate axial strength P_u is determined by specifying the eccentricity e in the axial load–deflection analysis. The ultimate moment capacity of the pin-ended beam-column is computed as $M_u = P_u \times e$. A set of ultimate axial strengths (P_u) and ultimate moment capacities (M_u) can be utilized to capture the load–moment strength envelopes. The pure ultimate axial strength P_{oa} of the column without bending moment is predicted by specifying the zero eccentricity in the axial load–deflection analysis. It is worth mentioning that the procedure for axial load–deflection analysis is not adequate to accurately calculate the pure bending moment M_o without the presence of the axial load. Therefore, the numerical procedure given by Liang (2011a) is employed to determine the axial load–moment strength diagrams for circular slender CFSST beam-columns.

An incremental and iterative analysis technique is used to model the complete axial load–moment strength diagrams for slender CFSST beam-columns. For the given axial load P_u, the ultimate moment capacity M_u is computed as the maximum moment $M_{e,max}$ that can be applied to the beam-column ends. The condition of moment equilibrium is maintained at the mid-height of the slender column. The external moment at the column mid-height is expressed mathematically by

$$M_{me} = M_e + P_u \left(u_o + u_m \right) \tag{3.7}$$

where M_e represents the applied moment at the column ends.

The mid-height deflection of the pin-ended slender CFSST beam-column can be obtained by means of rearranging Eq. (3.4) as

$$u_m = \phi_m \frac{L^2}{\pi^2} \tag{3.8}$$

For each mid-height curvature increment, the column end curvature is computed to achieve the moment equilibrium at the column mid-height by means of iterative numerical solution schemes. The residual moment generated at each iterative solution step is formulated as

$$r_{pm} = P_u\left(u_m + u_o\right) + M_e - M_{mi} \tag{3.9}$$

When the residual moment approaches to zero, the formulation assumes that the generalized stress equilibrium state is obtained. For this purpose, the convergence tolerance $\left|r_{pm}\right| < \varepsilon_k$ is specified as $\varepsilon_k = 10^{-4}$.

3.4.2 Modeling procedure for strength envelopes

The ultimate axial load of a CFSST slender column decreases with an increase in the loading eccentricity or bending moments at the column ends. To determine the maximum load increment in the generation of strength envelopes, the maximum axial load that the column can support must be computed, which is the pure ultimate axial load P_{oa} of the slender CFSST column under axial compression without any bending moments. For this purpose, the modeling procedure for the axial load–deflection responses described in Section 3.3.2 is employed. The applied axial load is then incrementally increased from zero to $0.9P_{oa}$ with a step of $0.1P_{oa}$. For the given axial load, the mid-height curvature ϕ_m of the column is gradually increased. For each mid-height curvature increment, the corresponding internal moment M_{mi} is determined by using the established axial load–moment–curvature relations. The column end curvature ϕ_e is iteratively adjusted to obtain the applied moment M_e at the column ends, which maintains the moment equilibrium condition at the column mid-height, until the maximum moment at the column ends $M_{e,\max}$ is obtained. The next axial load increment is applied, and the above process is repeated to determine the complete strength envelope. The modeling procedure for predicting the strength envelopes of slender columns proposed by Liang (2011a) is adopted and described as follows:

1. Input data.
2. Discretize the cross-section into fine fiber elements and calculate their coordinates and areas.

3. Calculate the pure ultimate axial load P_{oa} of the slender CFSST column by using the subroutine for axial load–deflection simulations.
4. Initialize the axial load $P_u = 0$.
5. Initialize the mid-height curvature of the slender column as $\phi_m = \Delta\phi_m$.
6. Compute the mid-height deflection u_m from the curvature ϕ_m at the column mid-height.
7. Calculate the internal section moment M_{mi} for the given axial load P_u using the established axial load–moment–curvature relationship.
8. Assign initial values to the curvature at the column ends: $\phi_{e,1} = 10^{-10}$, $\phi_{e,3} = 10^{-6}$, and $\phi_{e,2} = (\phi_{e,1} + \phi_{e,3})/2$.
9. Calculate the residual moments $r_{pm,1}$, $r_{pm,2}$, and $r_{pm,3}$ that correspond to $\phi_{e,1}$, $\phi_{e,2}$, and $\phi_{e,3}$, respectively.
10. Adjust the curvature at the column ends (ϕ_e) by using the numerical solution algorithms.
11. Calculate the moment M_e at the column ends based on the curvature ϕ_e by employing the subroutine for modeling the moment–curvature responses.
12. Repeat Steps (10)–(11) until $|r_{pm}| < \varepsilon_k$.
13. Increase the curvature at the column mid-height by applying $\phi_m = \phi_m + \Delta\phi_m$.
14. Repeat Steps (6)–(13) until the ultimate moment at the column ends $M_u \left(= M_{e,\max}\right)$ is obtained.
15. Increase the axial load by $P_u = P_u + 0.1P_{oa}$.
16. Repeat Steps (5)–(15) until the computation for the load increment of $0.9P_{oa}$ is completed.
17. Plot the axial load–moment strength envelope.

3.5 SOLUTION ALGORITHMS IMPLEMENTING MÜLLER'S METHOD

In the modeling of axial load–deflection responses, the neutral axis depth is iteratively adjusted to maintain the moment equilibrium at the column mid-length. The incremental equilibrium equation for the residual moment developed in an iterative numerical procedure is highly dynamic and non-linear, and nonderivative with respect to the design variables. The generalized displacement control method proposed by Yang and Shieh (1990) and Yang and Kuo (1994) can be used to solve this problem. Müller's approach (Müller 1956) is a numerical solution scheme for root finding of nonlinear equations and applicable to nonderivative nonlinear functions. Solution algorithms implementing Müller's approach have been developed to efficiently determine the true neutral axis depth. This approach starts with three initial neutral axis depths. The true neutral axis depth is determined by means of iteratively applying the following equations:

$$d_{n,4} = d_{n,3} - \frac{2r_{pu,3}}{q_1 \pm \sqrt{q_1^2 - 4p_1 r_{pu,3}}} \tag{3.10}$$

$$p_1 = \frac{\left(r_{pu,1} - r_{pu,3}\right)\left(d_{n,2} - d_{n,3}\right) - \left(r_{pu,2} - r_{pu,3}\right)\left(d_{n,1} - d_{n,3}\right)}{\left(d_{n,2} - d_{n,3}\right)\left(d_{n,1} - d_{n,2}\right)\left(d_{n,1} - d_{n,3}\right)} \tag{3.11}$$

$$q_1 = \frac{\left(r_{pu,2} - r_{pu,3}\right)\left(d_{n,1} - d_{n,3}\right)^2 - \left(r_{pu,1} - r_{pu,3}\right)\left(d_{n,2} - d_{n,3}\right)^2}{\left(d_{n,2} - d_{n,3}\right)\left(d_{n,1} - d_{n,2}\right)\left(d_{n,1} - d_{n,3}\right)} \tag{3.12}$$

The sign of square root in the denominator of Eq. (3.10) is taken to be the same as that of p_1. The values of $d_{n,1}$, $d_{n,2}$, and $d_{n,3}$ as well as the corresponding residual moments $r_{pu,1}$, $r_{pu,2}$, and $r_{pu,3}$ are exchanged to obtain the converged solutions.

In the modeling of column strength envelopes, both the neutral axis depth and curvature at the column ends need to be iteratively adjusted to meet the equilibrium conditions. The true curvature at the column ends is computed by means of utilizing the following expressions:

$$\phi_{e,4} = \phi_{e,3} - \frac{2r_{pm,3}}{q_2 + \sqrt{q_2^2 - 4p_2 r_{pm,3}}} \tag{3.13}$$

$$p_2 = \frac{\left(r_{pm,1} - r_{pm,3}\right)\left(\phi_{e,2} - \phi_{e,3}\right) - \left(r_{pm,2} - r_{pm,3}\right)\left(\phi_{e,1} - \phi_{e,3}\right)}{\left(\phi_{e,2} - \phi_{e,3}\right)\left(\phi_{e,1} - \phi_{e,2}\right)\left(\phi_{e,1} - \phi_{e,3}\right)} \tag{3.14}$$

$$q_2 = \frac{\left(r_{pm,2} - r_{pm,3}\right)\left(\phi_{e,1} - \phi_{e,3}\right)^2 - \left(r_{pm,1} - r_{pm,3}\right)\left(\phi_{e,2} - \phi_{e,3}\right)^2}{\left(\phi_{e,2} - \phi_{e,3}\right)\left(\phi_{e,1} - \phi_{e,2}\right)\left(\phi_{e,1} - \phi_{e,3}\right)} \tag{3.15}$$

The square root in the denominator of Eq. (3.13) is taken as positive to obtain the converged solution. The values of the curvature at the column ends $\phi_{e,1}$, $\phi_{e,2}$, and $\phi_{e,3}$, and the corresponding residual moments $r_{pm,1}$, $r_{pm,2}$, and $r_{pm,3}$ are exchanged to determine the converged solutions (Patel et al. 2012a, b).

3.6 ACCURACY OF MATHEMATICAL MODELS

3.6.1 Concentrically loaded columns

The mathematical modeling procedures developed are validated by comparing predicted ultimate axial strengths and axial load–deflection responses with experimental results of concentrically loaded circular slender CFSST columns reported by Uy et al. (2011). The dimensions as well as material properties of the tested specimens together with their ultimate axial strengths are presented in Table 3.1. The ratios of the computed ultimate axial strengths $(P_{u.\text{fib}})$ to the experimental ones $(P_{u.\exp})$ range from 0.90 to 1.05 with a mean of 0.96. Both the coefficient of variation and standard deviation are 0.060 and 0.062, respectively. The comparison indicates that the developed computer modeling technique is capable of accurately predicting the ultimate axial strengths of concentrically loaded circular slender CFSST columns. As depicted in Figure 3.2, the simulated axial load–deflection responses compare well with those obtained from the tests. The fiber-based mathematical model accurately captures the initial stiffness and the post-peak axial load–deflection responses of the tested columns.

Table 3.1 Comparison of ultimate axial strengths of axially loaded circular slender CFSST columns

Specimens	$D \times t$ (mm)	L (mm)	$\sigma_{0.2}$ (MPa)	E_0 (GPa)	n	f'_c (MPa)	$P_{u.\exp}$ (kN)	$P_{u.\text{fib}}$ (kN)	$\dfrac{P_{u.\text{fib}}}{P_{u.\exp}}$	$P_{u.\text{EC4}}$	$\dfrac{P_{u.\text{EC4}}}{P_{u.\exp}}$
C1-1a	113.6 × 2.8	485	288.6	173.9	7.6	36.3	738.0	760.24	1.04	699	0.95
C1-1b	113.6 × 2.8	485	288.6	173.9	7.6	75.4	1137.1	1126.97	0.99	1,026	0.90
C1-2a	113.6 × 2.8	1,540	288.6	173.9	7.6	36.3	578.9	595.73	1.03	544	0.94
C1-2b	113.6 × 2.8	1,540	288.6	173.9	7.6	75.4	851.1	827.97	0.97	805	0.95
C1-3a	113.6 × 2.8	2,940	288.6	173.9	7.6	36.3	357.6	374.37	1.05	340	0.95
C2-1a	101 × 1.48	440	320.6	184.2	7.2	36.3	501.3	456.77	0.91	465	0.93
C2-1b	101 × 1.48	440	320.6	184.2	7.2	75.4	819.0	748.85	0.91	737	0.90
C2-2a	101 × 1.48	1,340	320.6	184.2	7.2	36.3	446.0	403.02	0.90	372	0.84
C2-2b	101 × 1.48	1,340	320.6	184.2	7.2	75.4	692.9	634.20	0.92	582	0.84
C2-3b	101 × 1.48	2,540	320.6	184.2	7.2	75.4	389.7	355.39	0.92	278	0.71
Mean									0.96		0.89
Standard deviation (SD)									0.060		0.076
Coefficient of variation (COV)									0.062		0.086

Source: Adapted from Patel, V. I. et al., *Engineering Structures*, 130:1–13, 2017.

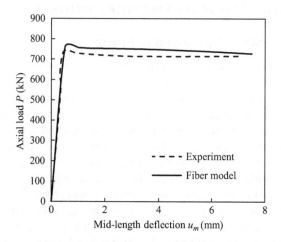

Figure 3.2 Comparison of load–deflection responses of Specimen Cl-la tested by Uy et al. (2011). (Reprinted from *Engineering Structures*, Vol. 130, Patel, V. I., Liang, Q. Q. and Hadi, M. N. S., Nonlinear analysis of circular high strength concrete-filled stainless steel tubular slender beam-columns, pp. 1–13, 2017. With permission.)

3.6.2 Eccentrically loaded columns

Experimental studies on circular slender CFSST beam-columns subjected to eccentric loads have not been reported in the published literature. Hence, experimental results on circular slender CFST beam-columns given by Portolés et al. (2011a) are employed to verify the mathematical model for circular slender CFSST beam-columns under eccentric loads. The material and geometric parameters of the tested columns are given in Table 3.2. It can be seen from Table 3.2 that the mathematical model accurately calculates the ultimate axial strengths of eccentrically loaded slender CFST columns. The statistical analysis shows that the mean ratio of the computational-to-experimental ultimate axial strength is 1.03 while the associated standard deviation and coefficient of variation are 0.073 and 0.071, respectively. The calculated and measured axial load–deflection responses of Specimen C100-5-3-30-20-1 are given in Figure 3.3. There is a good agreement between the predicted and measured results. However, it is observed that the experimental load–deflection curve slightly departs from the predicted one before attaining the ultimate axial strength. This discrepancy is likely caused by the uncertainty of the actual concrete compressive strength in the specimen. In addition, the mathematical simulation procedure accurately computes the post-peak responses of the tested slender CFST column.

Table 3.2 Comparison of ultimate axial strengths of eccentrically loaded circular slender CFST beam-columns

Specimens	$D \times t$ (mm)	L (mm)	e (mm)	f_y (MPa)	f_c' (MPa)	$P_{u.\exp}$ (kN)	$P_{u.\mathrm{fib}}$ (kN)	$\dfrac{P_{u.\mathrm{fib}}}{P_{u.\exp}}$
C100-3-2-30-20-1	100×3	2,135	20	322	32.70	181.56	210.70	1.16
C100-3-2-30-50-1	100×3	2,135	50	322	34.50	117.49	135.06	1.15
C100-3-2-70-20-1	100×3	2,135	20	322	65.79	248.58	247.52	1.00
C100-3-2-70-50-1	100×3	2,135	50	322	71.64	151.59	150.87	1.00
C100-3-2-90-20-1	100×3	2,135	20	322	95.63	271.04	272.89	1.01
C100-3-2-90-50-1	100×3	2,135	50	322	93.01	154.24	157.57	1.02
C100-3-3-30-20-1	100×3	3,135	20	322	39.43	140.32	150.28	1.07
C100-3-3-30-50-1	100×3	3,135	50	322	36.68	93.75	101.83	1.09
C100-3-3-70-20-1	100×3	3,135	20	322	71.74	159.55	166.26	1.04
C100-3-3-70-50-1	100×3	3,135	50	322	79.55	102.75	112.10	1.09
C100-3-3-90-20-1	100×3	3,135	20	322	94.56	160.33	175.34	1.09
C100-3-3-90-50-1	100×3	3,135	50	322	90.40	106.8	114.16	1.07
C100-5-2-30-20-1	100×3	2,135	20	322	35.39	270.02	288.72	1.07
C100-5-2-70-50-1	100×5	2,135	50	322	30.54	161.26	187.44	1.16
C100-5-2-70-20-1	100×5	2,135	20	322	70.16	313.55	324.80	1.04
C100-5-2-70-50-1	100×5	2,135	50	322	61.00	183.81	203.51	1.11
C100-5-2-90-20-1	101.6×5	2,135	20	320	95.43	330.4	363.55	1.10
C100-5-2-90-50-1	101.6×5	2,135	50	320	81.66	212.17	219.67	1.04
C100-5-3-30-20-1	101.6×5	3,135	20	320	38.67	212.48	215.24	1.01
C100-5-3-30-50-1	101.6×5	3,135	50	320	39.56	144.83	151.70	1.05
C100-5-3-70-20-1	101.6×5	3,135	20	320	71.89	231.35	232.31	1.00
C100-5-3-70-50-1	101.6×5	3,135	50	320	72.49	153.16	161.11	1.05
C100-5-3-90-20-1	101.6×5	3,135	20	320	86.39	246.82	239.03	0.97
C100-5-3-90-50-1	101.6×5	3,135	50	320	96.74	164.95	167.19	1.01
C125-5-3-90-20-1	125×5	3,135	20	322	87.98	474.17	435.06	0.92
C125-5-3-90-50-1	125×5	3,135	50	322	96.97	317.9	300.09	0.94
C125-5-3-90-20-2	125×5	3,135	20	322	107.33	489.47	459.50	0.94
C125-5-3-90-50-2	125×5	3,135	50	322	97.92	322.97	300.65	0.93
C160-6-3-90-20-1	160.1×5.7	3,135	20	322	87.38	1,012.5	923.13	0.91
C160-6-3-70-50-1	160.1×5.7	3,135	50	322	74.75	642.16	613.01	0.95
C160-6-3-90-20-2	160.1×5.7	3,135	20	322	83.08	1,011.5	910.11	0.90
Mean								1.03
Standard deviation (SD)								0.073
Coefficient of variation (COV)								0.071

Source: Adapted from Patel, V. I. et al., *Engineering Structures*, 130:1–13, 2017.

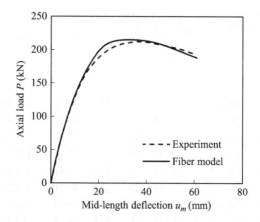

Figure 3.3 Comparison of load–deflection responses of Specimen C100-5-3-30-20-1 tested by Portolés et al. (2011a). (Reprinted from *Engineering Structures*, Vol. 130, Patel, V. I., Liang, Q. Q. and Hadi, M. N. S., Nonlinear analysis of circular high strength concrete-filled stainless steel tubular slender beam-columns, pp. 1–13, 2017. With permission.)

3.7 BEHAVIOR OF CIRCULAR SLENDER CFSST BEAM-COLUMNS

A computer program implementing the mathematical models has been developed and was employed to conduct parametric studies on the effects of geometric and material properties as well as concrete confinement on the structural behavior of circular slender CFSST beam-columns. In addition, the load distribution in the stainless steel tube and filled concrete in CFSST slender beam-columns were investigated. The details on the CFSST columns used in the parametric studies are provided in Table 3.3. The column initial imperfection of $L/1000$ was considered in computer simulations. The ultimate compressive strain of concrete was specified as 0.04 in the parametric studies.

3.7.1 Effects of column slenderness ratio

The column slenderness ratio (L/r) is an important geometric parameter that affects the fundamental behavior and failure modes of CFSST beam-columns. The mathematical modeling procedure was used to analyze eccentrically loaded CFSST beam-columns C1–C5 with various column slenderness ratios as given in Table 3.3. The computed axial load-deflection relations as a function of L/r ratios ranging from 22 to 100 are provided in Figure 3.4. As illustrated in Figure 3.4, the initial stiffness and ultimate

Table 3.3 Material and geometric parameters of circular slender CFSST beam-columns used in the parametric study

Columns	$D \times t$ (mm)	e/D	L/r	f'_c (MPa)	$\sigma_{0.2}$ (MPa)	E_0 (GPa)	n	$P_{u.fib}$ (kN)
C1	600 × 10	0.1	22	80	530	200	5	22,397
C2	600 × 10	0.1	30	80	530	200	5	20,873
C3	600 × 10	0.1	50	80	530	200	5	16,703
C4	600 × 10	0.1	70	80	530	200	5	12,726
C5	600 × 10	0.1	100	80	530	200	5	8,080
C6	500 × 10	0.1	30	65	530	200	5	14,138
C7	500 × 10	0.2	30	65	530	200	5	11,290
C8	500 × 10	0.3	30	65	530	200	5	9,154
C9	500 × 10	0.4	30	65	530	200	5	7,566
C10	500 × 10	0.5	30	65	530	200	5	6,384
C11	500 × 10	0.6	30	65	530	200	5	5,484
C12	700 × 14	0.1	30	100	320	200	7	28,563
C13	700 × 10	0.1	30	100	320	200	7	26,519
C14	700 × 7	0.1	30	100	320	200	7	24,810
C15	800 × 10	0.1	30	80	320	200	7	28,844
C16	800 × 10	0.1	30	80	530	200	5	33,481
C17	400 × 10	0.1	30	50	320	200	7	7,543
C18	400 × 10	0.1	30	80	320	200	7	9,432
C19	400 × 10	0.1	30	100	320	200	7	10,503
C20	550 × 10	0.1	50	120	320	200	7	15,678
C21	650 × 10	0.2	50	80	530	200	5	14,439

axial strengths of these columns decrease significantly as the slenderness L/r ratio increases. The reductions in the column ultimate axial load are 25%, 43%, and 64%, respectively, when increasing the L/r ratio from 22 to 50, 70, and 100. The smaller the L/r ratio, the steeper the post-peak branch of the axial load–deflection curve. The displacement–ductility and mid-height deflections at the ultimate axial loads remarkably increase with increasing the L/r ratio. This indicates that the second order has more significant effect on very slender columns rather than intermediate length columns. The strength envelopes obtained by the computational technique are presented in Figure 3.5. It is worth pointing out that the pure moment capacity is not affected by the L/r ratio. Nevertheless, the pure ultimate axial load and ultimate bending strength decreases significantly as the slenderness L/r ratio increases.

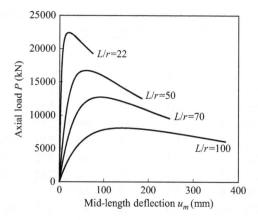

Figure 3.4 Effects of column slenderness ratio on the load–deflection behavior of circular CFSST beam-columns.

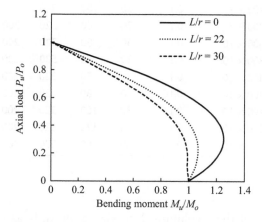

Figure 3.5 Effects of column slenderness ratio on the load–moment strength interaction diagrams of circular CFSST beam-columns.

3.7.2 Effects of load eccentricity ratio

The effects of load eccentricity ratio (e/D) on the performance of circular slender CFSST beam-columns were studied using the computer program. The Columns C6, C7, C9 and C11 given in Table 3.3 were simulated for this purpose. Figure 3.6 gives the load–deflection responses of circular slender CFSST beam-columns with various eccentricity ratios. It can be seen from Figure 3.6 that increasing the load eccentricity ratio significantly reduces the ultimate axial strength as well as initial stiffness of circular slender CFSST

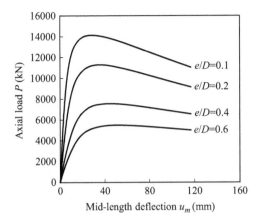

Figure 3.6 Effects of load eccentricity ratio on the load–deflection responses of CFSST slender columns.

beam-columns. This can be explained by the fact that the use of a larger e/D ratio results in larger bending moments at the column ends, which reduce the column ultimate axial load due to strength interaction. When the e/D ratio is increased from 0.1 to 0.2, 0.4, and 0.6, the reductions in the column ultimate axial strength are calculated to be 20%, 46%, and 61%, respectively. In addition, the larger the loading eccentricity ratio, the higher the displacement–ductility of the slender beam-columns. The effect of the e/D ratio on the column ultimate axial strengths is also demonstrated in Figure 3.7 by means of plotting the ultimate axial strength as a function of the eccentricity ratio.

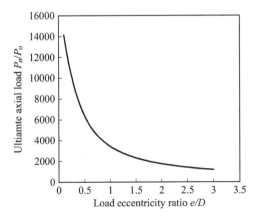

Figure 3.7 Effects of load eccentricity ratio on normalized ultimate axial loads of CFSST slender columns.

The computational results indicate that when the eccentricity ratios are large, the reduction in the column ultimate axial strength decreases as the e/D ratio increases. This is because the large e/D ratio significantly reduces the concrete confinement provided by the steel tube.

3.7.3 Effects of diameter-to-thickness ratio

The diameter-to-thickness ratio (D/t) of the stainless steel tube affects the characteristics of the circular slender CFSST beam-columns. Computational simulations on Columns C12, C13, and C14 given in Table 3.3 were undertaken using the mathematical model to examine the influences of the D/t ratio on the performance of circular slender CFSST beam-columns. The computed axial load–deflection responses of these columns are provided in Figure 3.8. It would appear that the D/t ratio has a minor influence on the initial stiffness of slender CFSST columns. However, it has a considerable effect on the ultimate axial loads of the CFSST slender columns. The column ultimate axial load decreases with increasing the D/t ratio or the tube thickness. When the D/t ratio is increased from 50 to 70 and 100, the reductions in the column ultimate load are found to be 7% and 13%, respectively. The strength envelopes of circular slender CFSST beam-columns with various D/t ratios are illustrated in Figure 3.9. The pure moment capacities are reduced by 25% and 46%, respectively, by changing the D/t ratio from 50 to 70 and 100.

3.7.4 Effects of stainless steel proof stress

Columns C15 and C16 listed in Table 3.3 were modeled by the computer program herein to investigate the effects of stainless steel proof stress $\sigma_{0.2}$ on

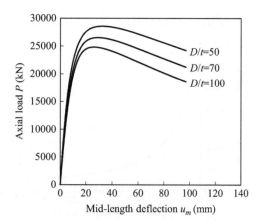

Figure 3.8 Effects of diameter-to-thickness ratio on load–deflection responses of slender CFSST columns.

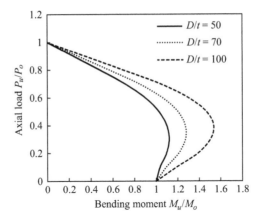

Figure 3.9 Effects of diameter-to-thickness ratio on load–moment diagrams of CFSST slender columns.

the behavior of circular slender CFSST beam-columns. Figure 3.10 presents the calculated axial load–deflection responses of circular slender CFSST beam-columns with proof stresses of 320 MPa and 530 MPa. It appears that the initial stiffness of slender beam-columns is not affected by the proof stress. However, the proof stress has a significant influence on the column strengths. The column ultimate axial load is found to increase by 16% when the stainless steel proof stress $\sigma_{0.2}$ is increased from 320 MPa to 530 MPa. The effect of stainless steel proof stress on the axial load–moment strength envelops is shown in Figure 3.11. It can be observed from Figure 3.11 that

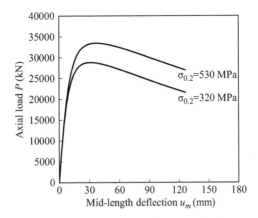

Figure 3.10 Effects of stainless steel proof stress on the load–deflection responses of slender CFSST columns.

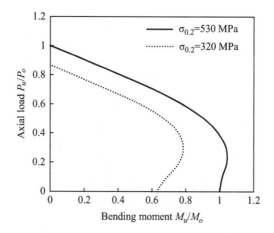

Figure 3.11 Effect of stainless steel proof stress on the load–moment strength envelopes of CFSST slender columns.

increasing the proof stresses from 320 MPa to 530 MPa results in a larger percentage increase in the pure moment capacity than in the pure ultimate axial strength of the slender CFSST column. This suggests that it is more beneficial to use higher strength stainless steel tube to increase the bending moment capacities of slender CFSST columns rather than to improve the column pure ultimate axial loads.

3.7.5 Effects of concrete compressive strength

The computer simulations on Columns C17, C18, and C19 given in Table 3.3 were performed to study the effects of concrete compressive strength f_c' on the responses of circular slender CFSST beam-columns. The influences of f_c' on the load–deflection responses of circular slender CFSST beam-columns under eccentric loading are demonstrated in Figure 3.12. The results show that the concrete strength has a minor influence on the initial stiffness of CFSST slender columns but has a significant influence on their ultimate axial strengths, and its effect decreases with an increase in the column slenderness as reported by Liang (2018). In particular, the ultimate axial strength of the slender CFSST column with 100 MPa concrete was about 28% higher than that of the one with 50 MPa concrete. The interaction strength envelopes for circular slender CFSST beam-columns constructed by concrete with different strengths are given in Figure 3.13. It is found that the axial resistance increases more than the moment capacity does when increasing the concrete strength. This suggests that higher strength concrete should be used to improve the ultimate axial loads of circular CFSST slender columns rather than their moment capacities.

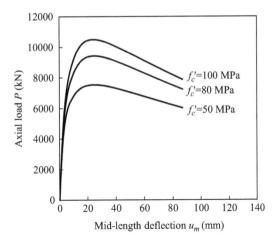

Figure 3.12 Effect of concrete strength on the load–deflection responses of CFSST slender columns.

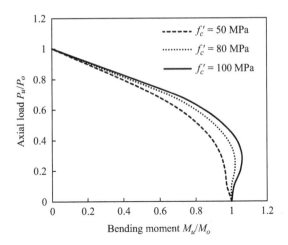

Figure 3.13 Effects of concrete strength on load–moment interaction diagrams of CFSST slender columns.

3.7.6 Effects of concrete confinement

The influences of confinement on the strength curves of circular CFSST slender columns subjected to eccentric loads have not been fully understood due to the lack of experimental and numerical studies. Therefore, it is important to investigate the effects of concrete confinement on the

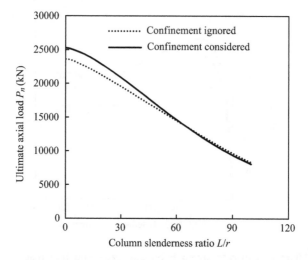

Figure 3.14 **Effects of concrete confinement on strength curves of circular CFSST beam-columns.**

structural performance of circular slender CFSST beam-columns. Column C1 given in Table 3.3 was analyzed by means of ignoring and considering the concrete confinement effect. The predicted column strength curves with or without consideration of confinement are provided in Figure 3.14. The influence of confinement on the column strength is shown to decrease with an increase in the column slenderness ratio. This effect is the most pronounced on the cross-sectional strength. The computer analysis ignoring concrete confinement effects underestimates the pure ultimate axial strength by 7%. It should be noted that concrete confinement does not increase the column strength when its slenderness ratio is greater than 70. Similar observations were reported by Liang (2011b) for circular slender CFST beam-columns. Therefore, the concrete confinement can be ignored in the analysis and design of CFSST beam-columns with the slenderness ratio greater than 70. The strength envelopes of the CFSST column having a slenderness of 22 are demonstrated in Figure 3.15. It is seen that the concrete confinement affects the pure ultimate axial strength of the beam-column more than its pure bending strength.

3.7.7 Load distribution in concrete and stainless steel tubes

The contributions of stainless steel tube and filled concrete to the ultimate strength of a circular slender CFSST beam-column can be evaluated

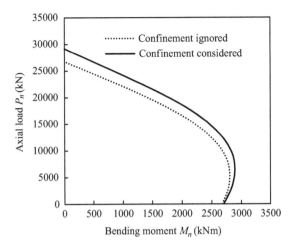

Figure 3.15 Effects of concrete confinement on the load–moment strength envelopes of circular CFSST beam-columns.

by means of contribution ratios. The contribution ratios are defined by Liang (2018) as

$$R_s = \frac{P_{u.s}}{P_u} \tag{3.16}$$

$$R_c = \frac{P_{u.c}}{P_u} \tag{3.17}$$

where R_s denotes the stainless steel contribution ratio, R_c is the concrete contribution ratio, and $P_{u.c}$ and $P_{u.s}$ are the axial loads carried by the concrete core and the stainless steel tube when the CFSST column attains its ultimate strength, respectively. The ultimate axial strength P_u of a slender CFSST beam-column is determined using the computational procedure for axial load–deflection responses.

The load distribution in steel tube and concrete in double-skin CFST slender columns subjected to eccentric loads has been investigated by Liang (2018). However, such work on circular slender CFSST columns has not been reported in the published literature. Therefore, the fiber model proposed was employed to study the load distribution in the stainless steel tube and concrete components in eccentrically loaded slender CFSST beam-columns. For this purpose, Column C20 given in Table 3.3 was analyzed to quantify the influences of column slenderness ratio (*L/r*) on the contribution ratios. Figures 3.16 and 3.17 present the predicted axial load–deflection curves of

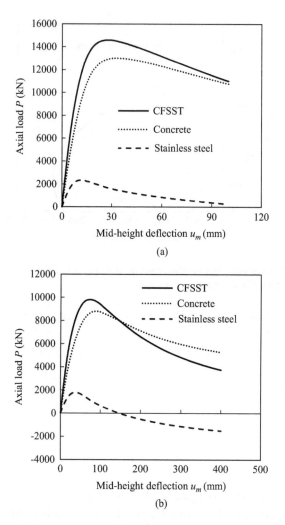

Figure 3.16 Load distribution in circular slender CFSST beam-columns: (a) L/r = 30 and (b) L/r = 60.

stainless steel tubes, concrete core, and CFSST beam-columns with various L/r ratios. It can be discovered that the concrete core carries a large portion of the axial load regardless of the L/r ratio. When the L/r ratio is less than or equal to 30, the stainless steel tube and concrete are subjected to compressive resultant forces. For the CFSST columns with L/r ratios ranging from 60 to 100, the stainless steel tubes are under tension when the columns are under large deflection. This is due to the fact that increasing the L/r ratio leads to an increase in the lateral deflection and bending moments at its mid-height. This results in tensile resultant force in the stainless steel

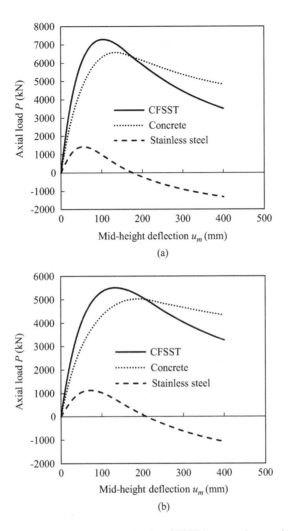

Figure 3.17 Load distribution in circular slender CFSST beam-columns: (a) $L/r = 80$ and (b) $L/r = 100$.

tubes. Figure 3.18 provides the stainless steel and concrete contribution ratios as a function of the L/r ratio. It appears that increasing the L/r ratio leads to an increase in the stainless steel contribution but a decrease in the concrete contribution. This implies that for CFSST columns with large L/r ratios, the use of high-strength concrete is not effective in increasing their ultimate axial strengths.

The numerical analysis on Column C21 given in Table 3.3 was conducted to evaluate the influences of eccentricity ratio (e/D) on the contribution ratios. The axial load–deflection responses of the stainless steel tube,

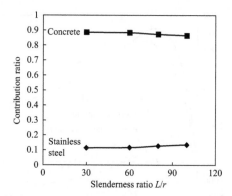

Figure 3.18 **Effects of column slenderness on the contribution ratios of stainless steel tube and concrete.**

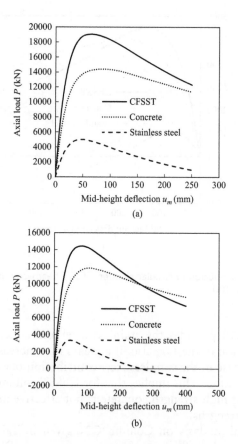

Figure 3.19 Load distribution in circular slender CFSST beam-columns: (a) $e/D = 0.1$ and (b) $e/D = 0.2$.

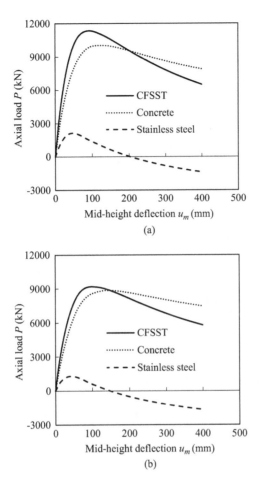

Figure 3.20 Load distribution in circular slender CFSST beam-columns: (a) $e/D = 0.3$ and (b) $e/D = 0.4$.

concrete core, and slender CFSST beam-columns with various e/D ratios are presented in Figures 3.19 and 3.20. The stainless steel tube and concrete are subjected to compressive resultant forces when the e/D ratio is less than or equal to 0.1. However, when the e/D ratio is greater than 0.1 and the column is under large deflection, the stainless steel tube is subjected to the tensile resultant force. The contribution ratios of the stainless steel tube and concrete in slender CFSST beam-columns as a function of the e/D ratio are shown in Figure 3.21. The figure indicates that the contribution of the concrete to the column ultimate strength is shown to increase with increasing the e/D ratio. In contrast, increasing the e/D ratio decreases the stainless steel contribution ratio.

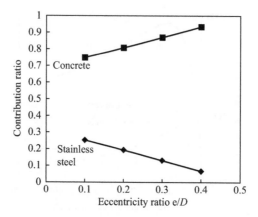

Figure 3.21 Effects of eccentricity ratio on the contribution ratios of stainless steel and concrete.

3.8 DESIGN OF CIRCULAR SLENDER CFSST COLUMNS

Eurocode 4 (2004) provides design equations for calculating the ultimate axial strength of circular CFST slender columns made of carbon steel tubes under axial compression. The codified methods include the effects of the column slenderness using column strength curves and concrete confinement. The design equations given in Eurocode 4 are modified by means of replacing the steel yield strength f_y with the stainless steel proof stress $\sigma_{0.2}$. The ultimate axial load of slender composite column under axial compression is given by

$$P_{u.EC4} = \chi N_{pl,Rd} \tag{3.18}$$

in which $N_{pl,Rd}$ represents the plastic resistance of the CFSST column section, which is determined by

$$N_{pl,Rd} = \eta_a A_s \sigma_{0.2} + A_c f_c' \left(1 + \eta_c \frac{t}{D} \frac{\sigma_{0.2}}{f_c'} \right) \tag{3.19}$$

where A_s and A_c are the cross-sectional area of the stainless steel tube and concrete, respectively, and η_a denotes the steel strength reduction factor which is given by

$$\eta_a = 0.25 \left(3 + 2\bar{\lambda} \right) \quad \left(\eta_a \leq 1.0 \right) \tag{3.20}$$

In Eq. (3.19), η_c stands for the concrete strength enhancement factor expressed by

$$\eta_c = 4.9 - 18.5\bar{\lambda} + 17\bar{\lambda}^2 \quad (\eta_c \geq 0) \tag{3.21}$$

It should be noted that η_c is equal to zero when $\bar{\lambda}$ is greater than 0.5. In Eq. (3.21), $\bar{\lambda}$ represents the column relative slenderness, calculated by

$$\bar{\lambda} = \sqrt{\frac{N_{pl,Rk}}{N_{cr}}} \tag{3.22}$$

where $N_{pl,Rk}$ denotes the characteristic compression plastic resistance of the composite cross-section, which is expressed by

$$N_{pl,Rk} = A_s\sigma_{0.2} + A_c f_c' \tag{3.23}$$

In Eq. (3.22), N_{cr} represents the elastic critical buckling load of the CFSST column, which is given by

$$N_{cr} = \frac{\pi^2 (EI)_{eff}}{L^2} \tag{3.24}$$

where $(EI)_{eff}$ denotes the effective flexural stiffness of a composite cross-section. The expression for $(EI)_{eff}$ is

$$(EI)_{eff} = E_0 I_s + 0.6 E_{cm} I_c \tag{3.25}$$

in which E_0 represents the Young's modulus of stainless steel, I_s and I_c are the second moments of area of the stainless steel tube and concrete, respectively, and E_{cm} represents the Young's modulus of concrete, which is given by

$$E_{cm} = 22,000 \left(\frac{f_c'+8}{10}\right)^{0.3} \tag{3.26}$$

In Eq. (3.18), χ represents the column buckling strength reduction factor which is expressed as

$$\chi = \frac{1}{\phi + \sqrt{\phi^2 - \bar{\lambda}^2}} \quad (\chi \leq 1.0) \tag{3.27}$$

Note that the factor χ is taken as unity if the slenderness $\bar{\lambda}$ is less than 0.2. In Eq. (3.27), ϕ is calculated by the following formula:

$$\phi = 0.5 \left[1 + \alpha_i \left(\bar{\lambda} - 0.2 \right) + \bar{\lambda}^2 \right] \tag{3.28}$$

where α_i denotes the imperfection factor which is taken as 0.21 as suggested by Liew and Xiong (2009).

The ultimate axial strengths of the test specimens were compared with those calculated by the method given in Eurocode 4 (2004) in Table 3.1. The average computed-to-test ratio of the ultimate axial strengths of circular slender CFSST columns is 0.89 while the corresponding coefficient of variation and standard deviation are 0.076 and 0.086, respectively. The comparison demonstrates that the design method provided in Eurocode 4 (2004) yields conservative predictions of the ultimate axial strengths of axially loaded circular CFSST slender columns. This is due to the fact that the approach provided in Eurocode 4 does not consider the beneficial effects of significant strain-hardening of stainless steel in the calculation of the column ultimate strengths.

3.9 CONCLUSIONS

The nonlinear analysis, behavior, and design of circular slender CFSST beam-columns have been described in this chapter. The fiber-based mathematical models for the simulations of circular slender CFSST beam-columns under eccentric loads have been presented. The numerical models take into consideration the effects of important features including the concrete confinement, column slenderness, loading eccentricity, initial imperfection, and material and geometric nonlinearities. Solution algorithms implementing Müller's numerical technique have been developed to solve the nonlinear dynamic equilibrium equations in an incremental iterative analysis procedure. The mathematical models proposed have shown to be an accurate and efficient computer simulation tool for predicting the load–deflection responses and load–moment interaction diagrams for circular slender CFSST beam-columns under eccentric loads.

The following important conclusions are drawn from the parametric study:

- The initial stiffness and ultimate axial loads of circular slender CFSST beam-columns are significantly reduced by means of increasing either the column slenderness or the loading eccentricity and are considerably reduced by increasing the diameter-to-thickness ratio.

- Increasing either the stainless steel proof stress or the concrete compressive strength remarkably increases the ultimate axial strengths of CFSST columns. However, their effects decrease with an increase in the column slenderness.
- The initial stiffness of slender CFSST columns is not affected by the stainless steel proof stress but is slightly influenced by the concrete compressive strength and diameter-to-thickness ratio.
- The displacement–ductility of CFSST slender columns is found to increase significantly by increasing the column slenderness or loading eccentricity, but increase slightly by increasing the diameter-to-thickness ratio.
- The effect of concrete confinement on the column axial strengths decreases with increasing the column slenderness and is most pronounced on the cross-sectional strength. Its effect can be ignored for CFSST columns with slenderness ratios greater than 70.
- Increasing the column slenderness ratio increases the contribution of the stainless steel tube to the column axial strength but decreases the concrete contribution.
- Increasing the loading eccentricity ratio decreases the contribution of the stainless steel tube to the column axial strength but increases the concrete contribution.
- It is more effective to use high-strength stainless steel tubes to improve the ultimate moment capacities of CFSST slender beam-columns rather than high-strength concrete. In contrast, high-strength concrete should be used to increase the ultimate axial strengths of CFSST slender beam-columns rather than their moment capacities.
- The design method specified in Eurocode 4 (2004) for CFST columns provides conservative predictions of the ultimate axial strengths of circular slender CFSST columns under axial compression.

REFERENCES

Abdella, K., Thannon, R. A., Mehri, A. I. and Alshaikh, F.A. (2011) "Inversion of three-stage stress-strain relation for stainless steel in tension and compression," *Journal of Constructional Steel Research*, 67(5): 826–832.

AISC 360-16 (2016). *Specification for Structural Steel Buildings*, Chicago, IL, USA: American Institute of Steel Construction.

AS/NZS 5100.6 (2017). *Australian/New Zealand Standard for Bridge Design, Part 6: Steel and Composite Construction*, Sydney, NSW, Australia: Standards Australia and Standards New Zealand.

Diniz, S. M. C. and Frangopol, D. M. (1997) "Strength and ductility simulation of high-strength concrete columns," *Journal of Structural Engineering*, ASCE, 123(10): 1365–1374.

Ellobody, E. (2013a) "A consistent nonlinear approach for analysing steel, cold-formed steel, stainless steel and composite columns at ambient and fire conditions," *Thin-Walled Structures*, 68: 1–17.

Ellobody, E. (2013b) "Nonlinear behaviour of eccentrically loaded FR concrete-filled stainless steel tubular columns," *Journal of Constructional Steel Research*, 90: 1–12.

El-Tawil, S., San-Picón, C. F., Deierlein, G. G. (1995) "Evaluation of ACI 318 and AISC (LRFD) strength provisions for composite beam-columns," *Journal of Constructional Steel Research*, 34(1): 103–123.

Eurocode 4 (2004) *Design of Composite Steel and Concrete Structures, Part 1.1: General Rules and Rules for Building*, BS EN 1994-1-1: 2004. London, UK: British Standards Institution.

Han, L. H., Xu, C. Y. Tao, Z. (in press) "Performance of concrete filled stainless steel tubular (CFSST) columns and joints: Summary of recent research," *Journal of Constructional Steel Research*. doi:10.1016/j.jcsr.2018.02.038

Hatzigeorgiou, G. D. (2008) "Numerical model for the behavior and capacity of circular CFT columns, Part I. Theory," *Engineering Structures*, 30(6): 1573–1578.

Knowles, R. B. and Park, R. (1969) "Strength of concrete-filled steel tubular columns," *Journal of the Structural Division*, ASCE, 95(12): 2565–2587.

Ky, V. S., Tangaramvong, S. and Thepchatri, T. (2015) "Inelastic analysis for the post-collapse behavior of concrete encased steel composite columns under axial compression," *Steel and Composite Structures*, 19(5): 1237–1258.

Liang, Q. Q. (2011a) "High strength circular concrete-filled steel tubular slender beam-columns, Part I: Numerical analysis," *Journal of Constructional Steel Research*, 67(2): 164–171.

Liang, Q. Q. (2011b) "High strength circular concrete-filled steel tubular slender beam-columns, Part II: Fundamental behavior," *Journal of Constructional Steel Research*, 67(2): 172–180.

Liang, Q. Q. (2014) *Analysis and Design of Steel and Composite Structures*, Boca Raton, FL, USA and London, UK: CRC Press, Taylor & Francis Group.

Liang, Q. Q. (2018) "Numerical simulation of high strength circular double-skin concrete-filled steel tubular slender columns," *Engineering Structures*, 168: 205–217.

Liew, J. Y. R. and Xiong, D. X. (2009) "Effect of preload on the axial capacity of concrete-filled composite columns," *Journal of Constructional Steel Research*, 65(3): 709–722.

Müller, D. E. (1956) "A method for solving algebraic equations using an automatic computer," *MTAC*, 10: 208–215.

Neogi, P. K., Sen, H. K., Chapman, J. C. (1969) "Concrete-filled tubular steel columns under eccentric loading," *The Structural Engineer*, 47(5): 187–195.

Patel, V. I., Liang, Q. Q., Hadi, M. N. S. (2012a) "High strength thin-walled rectangular concrete-filled steel tubular slender beam-columns. Part I: Modeling," *Journal of Constructional Steel Research*, 70: 377–384.

Patel, V. I., Liang, Q. Q., Hadi, M. N. S. (2012b) "High strength thin-walled rectangular concrete-filled steel tubular slender beam-columns. Part II: Behavior," *Journal of Constructional Steel Research*, 70: 368–376.

Patel, V. I., Liang, Q. Q., Hadi, M. N. S. (2017) "Nonlinear analysis of circular high strength concrete-filled stainless steel tubular slender beam-columns," *Engineering Structures*, 130: 1–13.

Portolés, J. M., Romero, M. L., Bonet, J. L., Filippou, F. C. (2011a) "Experimental study of high strength concrete-filled circular tubular columns under eccentric loading" *Journal of Constructional Steel Research*, 67(4): 623–633.

Portolés, J. M., Romero, M. L., Filippou, F. C., Bonet, J. L. (2011b) "Simulation and design recommendations of eccentrically loaded slender concrete-filled tubular columns," *Engineering Structures*, 33(5): 1576–1593.

Quach, W. M., Teng, J. G. and Chung, K. F. (2008) "Three-stage full-range stress-strain model for stainless steels," *Journal of Structural Engineering*, ASCE, 134(9): 1518–1527.

Rangan, B. and Joyce, M. (1992) "Strength of eccentrically loaded slender steel tubular columns filled with high-strength concrete," *Structural Journal*, ACI, 89(6): 676–681.

Rasmussen, K. J. R. (2003) "Full-range stress-strain curves for stainless steel alloys," *Journal of Constructional Steel Research*, 59(1): 47–61.

Shakir-Khalil, H., Zeghiche, J. (1989) "Experimental behaviour of concrete-filled rolled rectangular hollow-section columns," *The Structural Engineer*, 67(19): 346–353.

Shanmugam, N. E., Lakshmi, B., Uy, B. (2002) "An analytical model for thin-walled steel box columns with concrete in-fill," *Engineering Structures*, 24(6): 825–838.

Uy, B., Tao, Z., Han, L. H. (2011) "Behaviour of short and slender concrete-filled stainless steel tubular columns," *Journal of Constructional Steel Research*, 67(3): 360–378.

Yang, Y. B. and Kuo, S. R. (1994) *Theory and Analysis of Nonlinear Framed Structures*, New York, FL, USA and London, UK: Prentice Hall.

Yang, Y. B. and Shieh, M. S. (1990) "Solution method for nonlinear problems with multiple critical points," *AIAA Journal*, 28(12): 2110–2116.

Chapter 4

Nonlinear analysis of rectangular CFSST slender columns

4.1 INTRODUCTION

Rectangular and square slender concrete-filled stainless steel tubular (CFSST) beam-columns in composite buildings are often subjected to combined actions of axial load and bending moments. The CFSST columns at the corners of a composite frame are under combined axial load and biaxial bending. These bending moments are generally induced by the end moments from frame actions, loading eccentricities, and initial geometric imperfections. In practice, all columns must be designed as beam-columns, which are subjected to combined axial load and bending. Square and rectangular CFSST columns are widely used in high-rise composite buildings due to their ease of connections with steel beams compared to circular ones. To reduce the initial high costs of stainless steel material, thin-walled stainless steel tubes with relatively large width-to-thickness ratios are frequently used to construct rectangular CFSST columns. However, this gives rise to the outward local buckling of thin stainless steel tubes, which considerably reduces the column stiffness, strength, and ductility. Rectangular CFSST slender columns under biaxial loads may fail by the local and overall interaction buckling. The interaction buckling behavior of rectangular slender CFSST columns has not been fully understood owing to the lack of experimental and numerical studies on this challenging problem. This chapter addresses this problem in computer simulation aspects.

Experimental investigations on the responses of rectangular concrete-filled steel tubular (CFST) beam-columns under axial load and biaxial bending were carried out by various researchers, such as Bridge (1976), Shakir-Khalil and Zeghiche (1989), Wang (1999), Mursi and Uy (2004), Guo et al. (2011), Lai et al. (2014), Liew et al. (2016), and Xiong et al. (2017). Test results indicated that thin-walled CFST slender columns made of non-compact or slender steel sections failed by the interaction local and overall buckling. However, tests on rectangular slender CFSST beam-columns have been extremely scare. The first experiment on biaxially loaded square slender CFSST beam-columns constructed by steel fiber-reinforced concrete was conducted by Tokgoz (2015). It was observed that the member capacities of slender CFSST beam-columns were not affected by steel fibers. However, the addition of steel fibers in concrete increases the ductility of the CFSST beam-columns. Uy et al. (2011) conducted experiments on square and rectangular slender CFSST columns subjected to concentric axial compression. They reported that the tested CFSST columns failed by overall column buckling with large mid-height deflections.

Numerical models for determining the performance of slender CFST beam-columns subjected to axial load and biaxial bending have been developed by Lakshmi and Shanmugam (2002), Mursi and Uy (2006), Liang et al. (2012), Patel et al. (2015a, b), Xiong et al. (2017), and Liang (2018). However, only a few numerical studies on rectangular slender CFSST beam-columns under biaxial bending have been reported in the literature. Tokgoz (2015) presented a fiber model for the nonlinear analysis of square slender CFSST beam-columns subjected to axial load and biaxial bending. The fiber model assumed the same strain-hardening behavior of stainless steel in compression and tension and did not consider the local bucking of stainless steel tube under biaxial bending. Patel et al. (2017) developed an efficient numerical model using the fiber formulation for simulating the responses of rectangular CFSST slender beam-columns under biaxial loads. The numerical model recognized the different strain-hardening behaviors of stainless steel in compression and tension and was found to predict well the measured behavior of CFSST columns. A simple expression for calculating the pure moment capacity of square CFSST columns was proposed based on numerical studies by Patel et al. (2017).

This chapter describes theoretical models developed for predicting the local and overall interaction buckling behavior of rectangular CFSST slender beam-columns subjected to combined axial compression and biaxial bending. The formulations of the general theory and computational algorithms for determining the load–deflection responses and strength envelopes of rectangular CFSST slender beam-columns under axial load and biaxial bending are given. Theoretical models are used to investigate the fundamental behavior of CFSST slender columns. Important parameters are examined to study their sensitivities on the ultimate axial strength, concrete contribution ratio, pure moment capacity, load–deflection response, local buckling, applied load angle, cross-sectional shape, and the column strength curve. The design for bending of square CFSST columns and for the ultimate axial strength of concentrically loaded slender CFSST columns is discussed.

4.2 FORMULATION OF CROSS-SECTIONS UNDER BIAXIAL BENDING

The cross-section of a rectangular CFSST slender column under axial compression and biaxial bending are subjected to stress gradients as illustrated in Figure 2.8. Thin stainless steel tube walls under nonuniform compressive stresses may buckle locally away from the concrete core. The post-local buckling of thin stainless tube walls occurs continually as the applied biaxial loads gradually increase in a nonlinear inelastic analysis (Liang et al. 2007). The post-local buckling couples with the overall column buckling, which leads to the failure of the CFSST column. This failure process is also associated with the yielding of the stainless steel tube and crushing of the concrete infill. The simulation of a CFSST slender beam-column under a complete loading history must adequately capture the nonlinear inelastic cross-sectional characteristics, including the progressive post-local buckling, yielding, and cracking and crushing of the concrete core. The unique fiber modeling procedure for the post-local buckling of thin-walled steel tubes in concrete-filled composite columns under biaxial loads was proposed by Liang (2009a, b) and described in Chapter 2. This modeling scheme is adopted in the nonlinear analysis of CFSST slender beam-columns.

The axial load–moment–curvature relationships of cross-sections are required in the stability analysis of CFSST slender columns. These relationships can be generated by means of gradually increasing the curvature and solving for the corresponding section moment capacity as discussed in Chapter 2. For sections subjected to biaxial bending, the internal force and moments in both directions must be in equilibrium with external applied load and moments. To achieve the equilibrium conditions, the depth and orientation of the neutral axis must be iteratively adjusted using numerical solution algorithms. As discussed above, the computational analysis must account for inelastic behavior and higher order effects of post-local buckling in the calculation of axial force and moments.

4.3 SIMULATING LOAD–DEFLECTION RESPONSES FOR BIAXIAL BENDING

4.3.I General theory

The formulation of general theory for pin-ended rectangular thin-walled CFSST slender beam-columns under combined axial load and biaxial bending is presented in this section. The assumptions made for the formulation of the theoretical model are: (a) the slender beam-column is subjected to an equal load eccentricity at both ends; (b) the slender beam-column is under single curvature bending about its bending plane; (c) lateral deflection occurs in the bending plane of the beam-column; (d) the critical section is at mid-height of the beam-column. The deflection of a rectangular slender CFSST beam-column along its length is computed by means of using the part-sine shape function as

$$u = u_m \sin \frac{\pi z}{L} \qquad (4.1)$$

in which L and u_m represent the effective length and mid-height lateral deflection of the slender beam-column, respectively.

The curvature at the column mid-height in the plane of bending of the biaxially loaded slender CFSST beam-column can be obtained from its shape function as

$$\phi_m = u_m \frac{\pi^2}{L^2} \tag{4.2}$$

The second-order effect induced by the interaction of the applied axial load and lateral deflection amplifies the external bending moment at the column mid-height. Design codes require that the second-order effect must be considered in the determination of design actions on slender columns. The formulation of the theoretical model takes into account the initial geometric imperfection (u_o), loading eccentricity (e), and mid-height deflection (u_m), which represent the second-order effect. The external bending moment (M_{me}) at the column mid-height is therefore calculated by

$$M_{me} = P_a (u_o + u_m + e) \tag{4.3}$$

in which P_a stands for the axial load applied at a fixed angle α with respect to the y-axis and an eccentricity e as depicted in Figure 2.6.

To calculate the complete load–deflection curve consisting of ascending and descending branches, the analysis method of deflection control is employed in the theoretical formulation (Liang 2011a, b). The method calculates the applied axial load that corresponds to a given lateral deflection at the column mid-height (u_m). The complete load–deflection curve of the slender beam-column can be generated by means of repeating this process. As the critical section of the slender beam-column is at its mid-height, the equilibrium is maintained at the column mid-height. This condition requires that the external moment must be equal to the internal resultant moment in the plane of bending as well as the applied load angle (α) must be maintained. Therefore, for a slender CFSST column subjected to combined axial load and biaxial bending moments, the equilibrium equations are expressed in mathematical form as

$$M_{mi} - P(u_o + u_m + e) = 0 \tag{4.4}$$

$$\tan\alpha - \frac{M_y}{M_x} = 0 \tag{4.5}$$

in which the resultant internal moment $M_{mi} = \sqrt{M_x^2 + M_y^2}$, where M_x and M_y are the internal moments about the x-axis and y-axis, respectively. The internal moments are calculated as stress resultants by means of integrating the fiber stresses over the column cross-section about the principal axes, respectively, as described in Chapter 2.

The internal axial force P satisfying the equilibrium condition given by Eqs. (4.4) and (4.5) is determined as the applied axial load (P_a). This means that the external applied axial load is in equilibrium with the internal axial force. In a numerical analysis, equilibrium conditions are approximately satisfied by means of prescribing a convergence tolerance. The numerical analysis generates residual moment and moment ratio at each iteration, which are calculated as follows:

$$r_a = M_{mi} - P\left(u_o + u_m + e\right) \tag{4.6}$$

$$r_b = \tan\alpha - \frac{M_y}{M_x} \tag{4.7}$$

If the residual moment and moment ratio given above are small enough, the equilibrium is assumed to be maintained. This implies that if $|r_a| < \varepsilon_k$ and $|r_b| < \varepsilon_k$, in which $\varepsilon_k = 10^{-4}$ represents the prescribed convergence tolerance, the equilibrium condition is satisfied and the numerical solution converges.

4.3.2 Computer simulation procedure

The nonlinear inelastic load–deflection analysis of CFSST slender columns is generally an incremental and iterative process. The mid-height deflection (u_m) of the slender beam-column is incrementally increased. The curvature ϕ_m at the column mid-height is calculated from the corresponding deflection u_m. The internal axial force P and resultant internal moment M_{mi} are determined from the cross-section analysis. The depth d_n and orientation θ of the neutral axis are iteratively adjusted by means of using numerical solution algorithms to achieve the equilibrium conditions at the column mid-height. A computer program has been developed that calculates the load–deflection responses of rectangular CFSST slender beam-columns under biaxial loads considering the interaction local and overall column buckling. The computer program consists of a number of subroutines, which include Data, Fibers, Strains, Stresses, Forces, and Moments. The modeling of local and post-local buckling of rectangular stainless steel tube is included in the subroutine of Stresses.

The computer simulation procedure for the load–deflection responses of CFSST slender columns are described as follows (Liang et al. 2012):

1. The column cross-section is discretized into fine square fibers and their coordinates and areas are calculated.
2. The deflection at the column mid-height is initialized as $u_m = \Delta u_m$.
3. The curvature at the column mid-height is computed using Eq. (4.2).
4. The neutral axis depth d_n is iteratively adjusted using the numerical solution algorithms.

5. The internal axial force P and moment M_{mi} are calculated as stress resultants in the cross-section.
6. The residual moment r_a is determined.
7. The analysis procedure is repeated from Steps (4)–(6) until the condition $|r_a| < \varepsilon_k$ is satisfied.
8. The internal moments M_x and M_y are computed.
9. The orientation (θ) of the neutral axis is iteratively adjusted using the numerical solution algorithms.
10. The residual moment ratio (r_b) is computed.
11. The analysis procedure is repeated from Steps (4)–(10) until the condition $|r_b| < \varepsilon_k$ is attained.
12. The mid-height deflection is increased by $u_m = u_m + \Delta u_m$.
13. Steps (3)–(12) are repeated until the axial load is less than $0.65 P_{\max}$ or the prescribed deflection limit is exceeded.
14. The axial load–deflection curve is drawn.

4.4 MODELING STRENGTH ENVELOPES FOR BIAXIAL BENDING

4.4.1 Theoretical formulation

The axial load–moment strength envelopes can be employed to check the design capacities of rectangular slender CFSST beam-columns under axial load and biaxial bending. In the advanced analysis of composite frames, strength envelopes are used as yield surfaces that determine the limit state of composite columns. The formulation of the theoretical model considers the pin-ended CFSST slender beam-column illustrated in Figure 3.1. The axial load–moment strength envelope of the beam-column is generated by means of calculating the maximum ultimate moments $(M_{e.\max})$ that can be applied to the column ends for a set of ultimate axial loads (P_u) applied at a fixed angle (α). This means that the formulation determines the maximum unknown loading eccentricity at the column ends. This is different from the load–deflection analysis in which the loading eccentricity is known.

The theoretical model incorporates the effects of initial geometric imperfection and second order. The external bending moment at the column mid-length is the sum of the moment M_e at the column ends induced by the eccentric load, the moment $P_u u_o$ due to initial imperfection, and the second-order moment $P_u u_m$ and is written as

$$M_{me} = P_u (u_o + u_m) + M_e \tag{4.8}$$

The mid-height deflection u_m is computed from the curvature ϕ_m as follows:

$$u_m = \phi_m \frac{L^2}{\pi^2} \tag{4.9}$$

For an axial load increment, the theoretical modeling technique incrementally increases the curvature ϕ_m at the column mid-length and computes the corresponding internal moment capacity M_{mi} by employing the axial load–moment–curvature relationship. The moment at the column ends (M_e) is then determined by means of iteratively adjusting the curvature at the column ends (ϕ_e) until the moment at the column ends is maximized. The strength envelope can be determined by repeating the above computational process. The slender CFST beam-column subjected to axial load and biaxial bending must satisfy the following equilibrium equations as given by Liang et al. (2012):

$$P - P_u = 0 \tag{4.10}$$

$$M_{mi} - M_e - P_u \left(u_m + u_o \right) = 0 \tag{4.11}$$

$$\frac{M_y}{M_x} - \tan \alpha = 0 \tag{4.12}$$

The residual moment, moment ratio, and axial force generated at each iteration are calculated as

$$r_a = M_{mi} - P \left(e + u_m + u_o \right) \tag{4.13}$$

$$r_b = \frac{M_y}{M_x} - \tan \alpha \tag{4.14}$$

$$r_c = P - P_u \tag{4.15}$$

The residual moment, moment ratio, and axial force must be within the specified tolerance to obtain satisfactory solutions, such as $|r_a| < \varepsilon_k$, $|r_b| < \varepsilon_k$, $|r_c| < \varepsilon_k$, where $\varepsilon_k = 10^{-4}$.

4.4.2 Numerical modeling procedure

The computation of the strength envelope usually starts from the determination of the ultimate pure moment capacity in the absence of any axial load and incrementally increases the axial load to compute the corresponding ultimate moments. However, the ultimate pure axial load (P_{oa}) of the slender column without any moments need to be calculated first in order to determine appropriate load increments. This can be done by means of applying the analysis procedure for the axial load–deflection responses. Ten load increments are used in the modeling procedure to generate the complete axial load–moment strength envelope. The load step is set to $P_{oa}/10$.

The numerical modeling procedure for simulating the strength envelopes of rectangular slender CFSST beam-columns under axial load and biaxial bending is described as follows (Liang et al. 2012):

1. Input data on the geometry and material properties of the slender CFSST column.
2. The column cross-section is divided into square fine fiber elements and their coordinates and areas are calculated.
3. The ultimate axial load P_{oa} of the CFSST column under axial compression is computed by the computer simulation procedure for load–deflection responses.
4. The axial load is initialized as $P_u = 0$.
5. The curvature at the column mid-length ϕ_m is initialized as $\phi_m = \Delta\phi_m$.
6. The deflection at the column mid-length u_m is computed from the curvature ϕ_m.
7. The neutral axis depth d_n is adjusted using the numerical solution algorithms.
8. The internal axial force P is computed by the stress integration over the cross-section including local buckling effects.
9. The residual axial force r_c is calculated.
10. Steps (7)–(9) are repeated until the convergence criterion $|r_c| < \varepsilon_k$ is satisfied.
11. The internal section moments M_x and M_y are computed by considering local buckling effects.
12. The neutral axis orientation θ is adjusted using the numerical solution algorithms.
13. The calculation of the residual moment ratio r_b is carried out.
14. Steps (7)–(13) are repeated until the convergence criterion $|r_b| < \varepsilon_k$ is satisfied.
15. The internal resultant moment M_{mi} of the cross-section is determined.
16. The curvature ϕ_e at the column ends is adjusted by using numerical solution algorithms.
17. The determination of the moment M_e at the column ends is undertaken by including local buckling effects.
18. The residual moment r_a is calculated.
19. Steps (16)–(18) are repeated until the convergence criterion $|r_a| < \varepsilon_k$ is satisfied.
20. The curvature at the column mid-height ϕ_m is increased by $\phi_m = \phi_m + \Delta\phi_m$.
21. Steps (6)–(20) are repeated until the ultimate moment at the column ends M_e is maximized so that $M_u = M_{e.\max}$.
22. The axial load P_u is increased by applying $P = P + 0.1P_{oa}$.
23. Steps (5)–(22) are repeated until the maximum load increment $0.9P_{oa}$ is reached.
24. The axial load–moment strength envelope is plotted.

4.5 SOLUTION ALGORITHMS FOR COLUMNS UNDER BIAXIAL BENDING

In the simulation of the strength envelope for a slender beam-column under biaxial loads, the computation process iteratively adjusts the neutral axis depth and orientation as well as the curvature at the column ends to meet the equilibrium conditions of generalized stresses. The development of numerical solution algorithms that implements Müller's approach (Müller 1956) has been undertaken for the determination of the neutral axis depth and orientation as well as the curvature at the column ends (Liang et al. 2012). The depth and orientation of the neutral axis and the curvature are treated as design variables. The three initial values of each of the design variables are denoted as ω_1, ω_2, and ω_3. The corresponding residual moment, moment ratio, and axial force $r_{m,1}$, $r_{m,2}$, and $r_{m,3}$ are calculated by means of using the initialized values. The calculation of the new design variable ω_4 that approaches the true value is undertaken by means of iteratively executing the following equations (Liang et al. 2012):

$$\omega_4 = \omega_3 - \frac{2r_{m,3}}{b_m \pm \sqrt{b_m^2 - 4a_m r_{m,3}}} \tag{4.16}$$

$$a_m = \frac{(r_{m,1} - r_{m,3})(\omega_2 - \omega_3) - (r_{m,2} - r_{m,3})(\omega_1 - \omega_3)}{(\omega_2 - \omega_3)(\omega_1 - \omega_3)(\omega_1 - \omega_2)} \tag{4.17}$$

$$b_m = \frac{(r_{m,2} - r_{m,3})(\omega_1 - \omega_3)^2 - (r_{m,1} - r_{m,3})(\omega_2 - \omega_3)^2}{(\omega_2 - \omega_3)(\omega_1 - \omega_3)(\omega_1 - \omega_2)} \tag{4.18}$$

The sign of b_m and square root in Eq. (4.16) is the same. The values of ω_1, ω_2, and ω_3 and the corresponding residual forces and moments $r_{m,1}$, $r_{m,2}$, and $r_{m,3}$ need to be exchanged as discussed by Patel et al. (2012). The values $D/4$, D, and $(d_{n,1} + d_{n,3})/2$ are given to the neutral axis depths $d_{n,1}$, $d_{n,3}$, and $d_{n,2}$, respectively. The initial neutral axis orientations θ_1, θ_3, and θ_2 are taken as $\alpha/4$, α, and $(\theta_1 + \theta_3)/2$, respectively. The column end curvatures $\phi_{e,1}$, $\phi_{e,2}$, and $\phi_{e,3}$ are assigned as 10^{-10}, 10^{-6}, and $(\phi_{e,1} + \phi_{e,3})/2$, respectively.

4.6 VERIFICATION OF THEORETICAL MODELS

4.6.1 Columns under axial loading

Experimental results on square slender CFSST columns under axial compression presented by Uy et al. (2011) were used to validate the mathematical models developed. Table 4.1 provides details on the 12 CFSST slender columns with the depth-to-thickness ratios (D/t) of 36 and 52 tested by Uy et al. (2011).

These columns were constructed by austenitic stainless steel tubes with proof stresses of 363 MPa or 390 MPa and filled with either normal strength concrete of 36 MPa or high-strength concrete of 75 MPa. The local buckling of the stainless steel tubes was taken into account in the nonlinear fiber analysis of these columns. The comparison of the ultimate axial strengths obtained from experiments and fiber analysis is given in Table 4.1. The predicted-to-tested ultimate axial strength $\left(P_{u.\text{fib}}/P_{u.\text{exp}}\right)$ ratios are between 0.85 and 1.02 with a mean value of 0.94. The coefficient of variation and standard deviation are 0.046 and 0.049, respectively. It appears that the developed fiber-based theoretical model is capable of accurately calculating the ultimate axial strengths of rectangular slender CFSST columns under axial compression.

The predicted axial load–deflection curve for Specimen S1-2b is compared with the experimentally measured one in Figure 4.1. Good agreement between numerical predictions and test results is obtained. The measured initial stiffness of Specimen S1-2b agrees well with the computed result. The measured post-peak branch of the load–deflection curve slightly departs from the predicted one. This discrepancy can be explained by the fact that the actual strength of concrete in the specimen was unknown and the average compressive strength of concrete cylinders was used in the fiber analysis. It can be concluded that the theoretical model accurately predicts the axial load–deflection responses of axially loaded rectangular slender CFSST columns.

4.6.2 Beam-columns under axial load and biaxial bending

The theoretical model was employed to compute the load–deflection curves of biaxially loaded square slender CFSST beam-column tested by Tokgoz (2015). The dimensions of Specimen CFSSTC-III were $100 \times 100 \times 3$ mm with a length of 1200 mm. The column was subjected to axial load and biaxial bending with an eccentricity of 65 mm in both the x and y directions. The proof stress of the stainless steel tube was 650 MPa, while its Young's modulus was 200 GPa. The compressive strength of the filled concrete was 58.42 MPa. It should be noted that the initial imperfection of the tested beam-column was not measured. The initial imperfection of $L/1000$ was assumed in the inelastic stability analysis. The load was not applied up to the column ultimate axial strength in the test. Therefore, only the predicted initial stiffness of the tested column was compared with the experimental one in Figure 4.2. The figure shows that excellent agreement between the predicted and measured load–deflection responses under axial load up to 200 kN is obtained. It is shown that the theoretical model is capable of accurately determining the performance of rectangular slender CFSST beam-columns subjected to axial load and biaxial bending. It is worth pointing out that the theoretical model developed was also verified by Patel et al. (2015a) for biaxially loaded rectangular CFST beam-columns for which the complete load–deflection responses were accurately captured.

Table 4.1 Ultimate axial strengths of rectangular slender CFSST beam-columns under axial compression

Specimens	$B \times D \times t$ (mm)	L (mm)	$\sigma_{0.2}$ (MPa)	E_0 (GPa)	n	f'_c (MPa)	$P_{u.exp}$ (kN)	$P_{u.fib}$ (kN)	$P_{u.EC4}$ (kN)	$\dfrac{P_{u.fib}}{P_{u.exp}}$	$\dfrac{P_{u.EC4}}{P_{u.exp}}$
S1-1a	100.3 × 100.3 × 2.76	440	390.3	182.0	6.7	36.3	767.6	696.4	746	0.91	0.97
S1-1b	100.3 × 100.3 × 2.76	440	390.3	182.0	6.7	75.4	1090.5	1046.4	1095	0.96	1.00
S1-2a	100.3 × 100.3 × 2.76	1340	390.3	182.0	6.7	36.3	697.3	657.5	678	0.94	0.97
S1-2b	100.3 × 100.3 × 2.76	1340	390.3	182.0	6.7	75.4	1022.9	987.5	957	0.97	0.94
S1-3a	100.3 × 100.3 × 2.76	2540	390.3	182.0	6.7	36.3	622.9	529.5	475	0.85	0.76
S1-3b	100.3 × 100.3 × 2.76	2540	390.3	182.0	6.7	75.4	684.2	700.7	565	1.02	0.83
R1-1a	49.0 × 99.5 × 1.93	440	363.3	195.3	6.1	36.3	385.6	339.9	360	0.88	0.93
R1-1b	49.0 × 99.5 × 1.93	440	363.3	195.3	6.1	75.4	558.3	508.1	527	0.91	0.94
R1-2a	49.0 × 99.5 × 1.93	740	363.3	195.3	6.1	36.3	361.1	335.4	351	0.93	0.97
R1-2b	49.0 × 99.5 × 1.93	740	363.3	195.3	6.1	75.4	517.7	500.1	509	0.97	0.98
R1-3a	49.0 × 99.5 × 1.93	2540	363.3	195.3	6.1	36.3	262.8	247.7	225	0.94	0.86
R1-3b	49.0 × 99.5 × 1.93	2540	363.3	195.3	6.1	75.4	332.8	322.6	267	0.97	0.80
Mean										0.94	0.91
Standard deviation (SD)										0.046	0.079
Coefficient of variation (COV)										0.049	0.089

Source: Adapted from Patel,V. I. et al., *Engineering Structures*, 140:120–133, 2017.

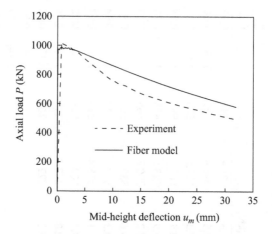

Figure 4.1   Comparison of axial load–deflection curves for Specimen S1-2b. (Reprinted from *Engineering Structures*, Vol. 140, Patel, V. I., Liang, Q. Q. and Hadi, M. N. S., Nonlinear analysis of biaxially loaded rectangular concrete-filled stainless steel tubular slender beam-columns, pp. 120–133, 2017. With permission.)

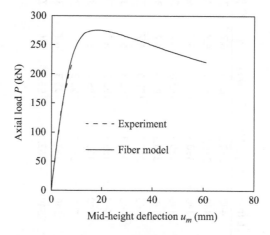

Figure 4.2   Comparison of axial load–deflection curves for Specimen CFSSTC-III 100×100×3. (Reprinted from *Engineering Structures*, Vol. 140, Patel, V. I., Liang, Q. Q. and Hadi, M. N. S., Nonlinear analysis of biaxially loaded rectangular concrete-filled stainless steel tubular slender beam-columns, pp. 120–133, 2017. With permission.)

4.7 BEHAVIOR OF RECTANGULAR SLENDER CFSST BEAM-COLUMNS

The computer program implementing the theoretical models was employed to examine the effects of depth-to-thickness ratio, column slenderness ratio, eccentricity ratio, applied angle, stainless steel proof stress, concrete

Table 4.2 Geometric and material properties of CFSST columns employed in the parametric study

Specimens	B×D×t (mm)	e/D	α (°)	L/r	f'_c (MPa)	$\sigma_{0.2}$ (MPa)	E_0 (GPa)	n
R1	400×500×5	0.1	30	100	40	490	212	6
R2	500×500×5	0.1	45	100	50	390	182	7
R3	300×300×3	0.1	30	22	80	360	195	6
R4	550×550×5.5	0.1	60	40	65	340	192	6
R5	700×700×7	0.1	60	22	65	360	195	6
R6	700×700×35	0.1	45	22	100	360	195	6
R7	550×550×10	0.1	45	30	80	390	182	7
R8	450×450×10	0.1	60	22	65	340	192	6
R9	300×400×10	0.1	45	30	80	340	192	6

compressive strength, and local buckling on the structural behavior of biaxially loaded rectangular slender CFSST beam-columns. The dimensions and properties of CFSST columns used in the parametric studies are given in Table 4.2. All numerical analyses accounted for the initial geometric imperfection of $L/1000$ at the column mid-length. The ultimate compressive strain of concrete (ε_{cu}) was taken as 0.04 in the numerical analyses.

4.7.1 Ultimate axial strengths

The influences of the column slenderness ratio (L/r), depth-to-thickness ratio (D/t), loading eccentricity ratio (e/D), and concrete compressive strength (f'_c) on the ultimate axial strengths of rectangular slender CFSST beam-columns were investigated. The beam-column R1 given in Table 4.2 was analyzed. Figure 4.3 demonstrates the effects of the above-mentioned factors on the column ultimate axial strengths. It appears from Figure 4.3 that the column ultimate axial strength decreases as the depth-to-thickness ratio or load eccentricity ratio increases. This is due to the fact that the stainless steel tube with a large depth-to-thickness ratio undergoes local buckling. The influence of loading eccentricity ratio emphasizes the fact that both primary moment $(P \cdot e)$ and secondary moment $(P \cdot u_m)$ increase as the loading eccentricity ratio increases. In contrast to the influences of the loading eccentricity ratio and depth-to-thickness ratio, the ultimate axial strength increases as the concrete compressive strength increases. Increasing the column slenderness ratio significantly decreases the ultimate strength of rectangular CFSST beam-columns. As shown in Figure 4.3, for columns with an e/D ratio greater than 2, the effects of concrete compressive strength, depth-to-thickness ratio, and column slenderness ratio become insignificant.

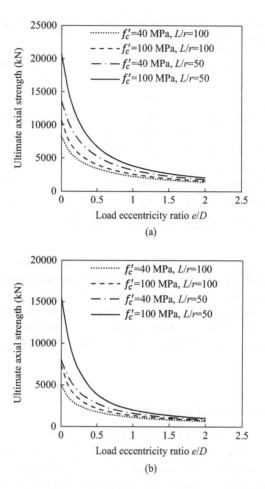

Figure 4.3 Effects of geometric and material properties on the ultimate axial strengths: (a) $D/t = 40$ and (b) $D/t = 100$.

4.7.2 Concrete contribution ratio

The contribution of concrete to the ultimate axial load (P_u) of slender CFSST beam-columns can be evaluated by the concrete contribution ratio, which is defined as $\xi_c = (P_u - P_s)/P_u$, where P_u is the ultimate axial strength of a slender CFSST beam-column under biaxial loads and P_s represents the ultimate axial strength of hollow stainless steel tubular beam-column (Liang et al. 2012). The fiber element analyses on the beam-column R2 given in Table 4.2 with various D/t ratios, concrete compressive strengths, e/D ratios, and L/r ratios were performed using the computer program. The computed concrete contribution ratios of the CFSST columns are given in Figure 4.4. The concrete contribution ratio (ξ_c) decreases as the loading eccentricity ratio or column slenderness ratio increases for the same depth-to-thickness ratio. The higher

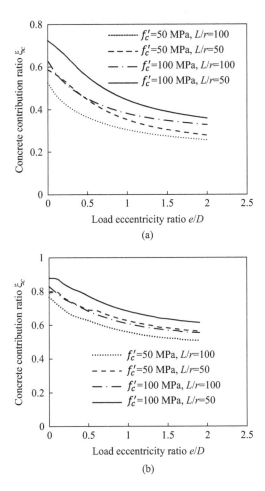

Figure 4.4 Effects of geometric and material properties on the concrete contribution ratio: (a) D/t = 50 and (b) D/t = 100.

the concrete strength, the larger the concrete contribution ratio, provided that the other parameters of the column are not changed. This suggests that when the CFSST column is relatively short with a small loading eccentricity, high-strength concrete can be used to construct these columns to achieve economical designs. However, the beneficial effect of high-strength concrete diminishes with increasing the column slenderness or loading eccentricity.

4.7.3 Pure moment capacities

The theoretical model was used to examine the influences of depth-to-thickness ratio (D/t), concrete strength (f_c'), and stainless steel proof strength $(\sigma_{0.2})$ on the ultimate pure moment capacity (M_o) of square

CFSST columns. Column R3 given in Table 4.2 was analyzed. The influences of the depth-to-thickness ratio on the ultimate pure moment capacity are illustrated in Figure 4.5, where Z_e is the elastic section modulus. As shown in Figure 4.5, increasing the depth-to-thickness ratio significantly decreases the ratio of $M_o/(Z_e\sigma_{0.2})$. This can be explained by the decrease in the cross-sectional area of stainless steel tube with the increase in the depth-to-thickness ratio. The effects of concrete compressive strength f_c' on the ratio of $M_o/(Z_e\lambda_m\sigma_{0.2})$ are shown in Figure 4.6, in which the factor γ_m depends on the D/t ratio and is given in Section 4.8.1. It can be seen from Figure 4.6 that increasing the concrete strength considerably increases the

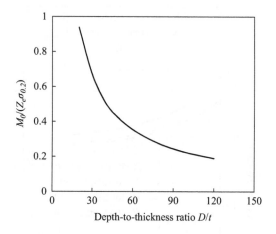

Figure 4.5   **Effects of depth-to-thickness ratio on the ratio of $M_o/(Z_e\sigma_{0.2})$.**

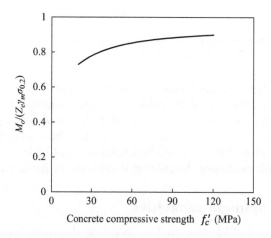

Figure 4.6   **Effects of concrete compressive strength f_c' on the ratio of $M_o/(Z_e\lambda_m\sigma_{0.2})$.**

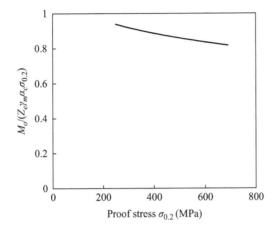

Figure 4.7   Effects of stainless steel proof stress $\sigma_{0.2}$ on the ratio of $M_o/(Z_e \lambda_m \alpha_c \sigma_{0.2})$.

ratio of $M_o/(Z_e\lambda_m\sigma_{0.2})$. Figure 4.7 demonstrates the influences of stainless steel proof strength $\sigma_{0.2}$ on the ratio of $M_o/(Z_e\lambda_m\alpha_c\sigma_{0.2})$, where α_c is a function of the concrete compressive strength and is given in Section 4.8.1. It appears that the ratio of $M_o/(Z_e\lambda_m\alpha_c\sigma_{0.2})$ decreases considerably with an increase in the stainless steel proof stress $\sigma_{0.2}$.

4.7.4 Axial load–deflection responses

The influences of geometric and material properties on the load–deflection behavior of biaxially loaded square slender CFSST beam-columns were investigated using the fiber-based theoretical model. Numerical analyses on the beam-column R4 given in Table 4.2 were carried out. The axial load–deflection responses of CFSST beam-columns with various parameters are given in Figures 4.8–4.12. As shown in Figure 4.8, increasing the D/t ratio notably reduces the column initial stiffness and considerably decreases the column ultimate axial strength. When increasing the D/t ratio from 50 to 70 and 100, the ultimate axial strength of the slender beam-column reduces by 12% and 21%, respectively. As illustrated in Figures 4.9 and 4.10, however, increasing the eccentricity ratio e/D or the column slenderness ratio L/r significantly reduces the column initial stiffness and ultimate axial strength. When increasing the e/D ratio from 0.2 to 0.4 and 0.6, the column ultimate axial strength reduces by 38% and 58%, respectively. The column ultimate axial load decreases by 27% and 56% when the L/r ratio is increased from 22 to 60 and 100, respectively. The column ultimate axial load increases significantly with increasing either the concrete compressive strength or the stainless steel tube proof stress. However, the use of higher strength material results in a slight improvement in the column initial stiffness. When increasing the concrete strength from 50 MPa to 80 MPa and 100 MPa, the ultimate axial strength increases by 42% and 69%, respectively. When increasing the stainless steel proof stress from 240 MPa to 530 MPa, the ultimate axial strength of the slender beam-column increases by 14%. In contrast to the effects of material strengths and geometric parameters on the initial stiffness and ultimate axial strength, the column displacement ductility generally decreases with an increase in material strengths but increases as the aforementioned geometric parameters increase.

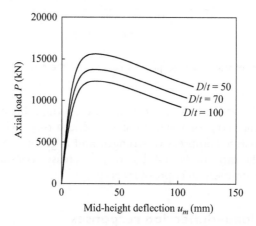

Figure 4.8   Effects of the depth-to-thickness ratio on the load–deflection responses of square CFSST slender beam-columns.

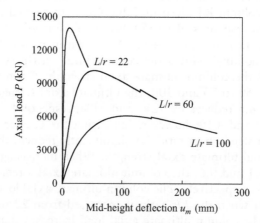

Figure 4.9   Effects of column slenderness ratio on the load–deflection responses of square CFSST slender beam-columns.

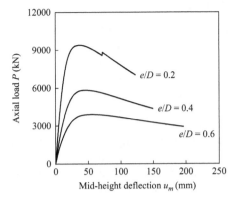

Figure 4.10   Effects of end eccentricity ratio on the load–deflection responses of square CFSST slender beam-columns.

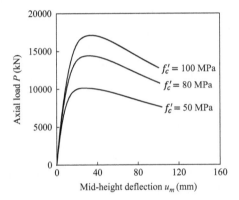

Figure 4.11   Effects of concrete strengths on the load–deflection responses of square CFSST slender beam-columns.

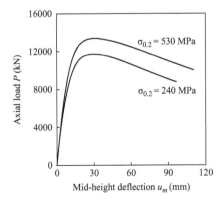

Figure 4.12   Effects of proof stress on the load–deflection responses of square CFSST slender beam-columns.

4.7.5 Local buckling

The local buckling of stainless steel walls significantly affects the behavior of thin-walled rectangular CFST beam-columns (Liang 2009a, b). However, numerical studies on the influences of local buckling on the ultimate strengths of biaxially loaded rectangular slender CFSST beam-columns have been very limited. The influences of the local buckling on the ultimate strengths of CFSST beam-columns were examined using the fiber-based theoretical model. The beam-column R5 given in Table 4.2 was analyzed by means of including or excluding local buckling effects, respectively. The load–deflection responses of the CFSST slender column are given in Figure 4.13. It is shown that the column ultimate axial load is overestimated by ignoring local buckling in the analysis. The normalized column strength $\left(P_u/P_{oe}\right)$ curves for the beam-column R5 given in Table 4.2 are illustrated in Figure 4.14, where P_{oe} denotes the ultimate strength of the eccentrically loaded composite cross-section. As shown in Figure 4.14, the influence of local buckling on the ultimate strength becomes more significant when the column slenderness ratio L/r decreases to zero. The effect of local buckling on the column strength decreases with increasing the column slenderness ratio L/r. The reduction in the column ultimate axial strength with zero length due to local buckling is about 10%. However, for a beam-column having an L/r ratio of 200, the local buckling reduces the column ultimate strength by only 0.7%.

Figure 4.15 shows a comparison of axial load–moment strength envelopes with and without considering local buckling effects. The ultimate axial strength with zero bending moment decreases generally more than the pure moment capacity due to the local buckling of stainless steel tube. The local buckling reduces the ultimate axial strength of the column

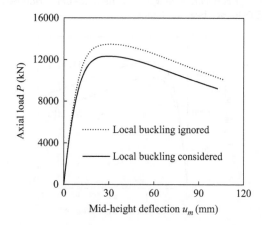

Figure 4.13   Effects of local buckling on the load–deflection responses of the square CFSST beam-column.

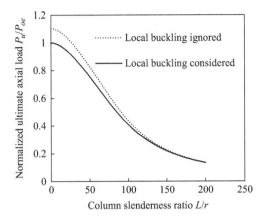

Figure 4.14   Effects of local buckling on the column strength curves of the square CFSST beam-column.

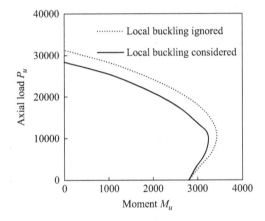

Figure 4.15   Effects on local buckling on the axial load–moment strength envelope of the square CFSST slender column.

without bending moments by 10% but has only a minor effect on the ultimate bending strength in the absence of axial load. This indicates that the effects of local buckling must be considered in the prediction of the axial load–moment strength envelopes. However, for very slender CFSST beam-columns, local buckling has only minor effects on their ultimate pure moment capacities.

4.7.6 Applied load angles

The theoretical model was used to investigate the influences of the applied load angle on the ultimate strengths of CFSST beam-columns. The applied

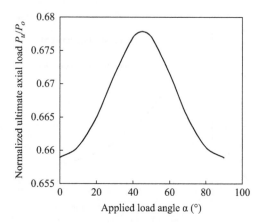

Figure 4.16   Effects of the applied load angle on the ultimate strength of the square CFSST beam-column.

load angle was varied from 0° to 90°. The beam-column R6 given in Table 4.2 was analyzed by means of changing the loading angle. The effects of loading angle α on the ultimate axial strength of square slender CFSST beam-columns are demonstrated in Figure 4.16, in which P_o is the ultimate axial strength of the column cross-section under axial compression. It can be observed from Figure 4.16 that when the loading angle is equal to 0° or 90°, the column ultimate axial strength is the lowest. However, the column has the maximum ultimate axial strength when the loading angle is equal to 45°. It is interesting to note that the column ultimate axial strength increases as the loading angle increases from 0° to 45° but decreases when the loading angle increases from 45° to 90°. It appears that the effects of applied load angle on the ultimate axial strength is not significant as shown in Figure 4.16.

4.7.7 Cross-sectional shapes

The effects of cross-sectional shape on the structural behavior of slender CFSST beam-columns were examined by means of conducting nonlinear analyses on both square and circular slender CFSST beam-columns. Both circular and square columns had the same cross-section area, column slenderness ratio (L/r), depth-to-thickness ratio (D/t), and material properties. The square beam-column R7 given in Table 4.2 was used. The diameter of the circular section was calculated as 620.21 mm. The predicted axial load–deflection responses of both columns are presented in Figure 4.17. It is seen from Figure 4.17 that the circular column has slightly higher initial stiffness than the square one. In addition, the circular column has higher ultimate axial strength than the square one. The ultimate axial strength of circular slender CFSST beam-column is 7% higher than that of the square

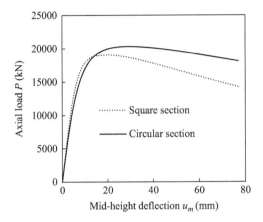

Figure 4.17   Effects of sectional shape on the axial load–deflection curves of slender square CFSST beam-columns.

one. In addition, the circular column displays better ductility than the square one. This is because circular stainless steel tube with a D/t ratio of 55 provides confinement to the filled concrete, which increases the strength and ductility of the filled concrete. On the other hand, the square steel tube with a D/t ratio of 55 undergoes local buckling, which considerably reduces its strength and ductility.

The axial load–moment strength interaction diagrams for circular and square slender CFSST beam-columns are given in Figure 4.18. It can be seen from the figure that the sectional shape has a pronounced effect on the column ultimate axial load. However, its effect diminishes with an

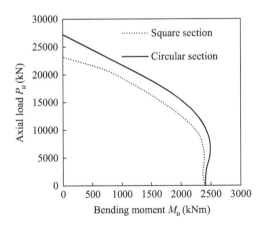

Figure 4.18   Effects of sectional shape on the axial load–moment interaction diagram for slender CFSST beam-columns.

increase in the bending moment. The ultimate pure axial strength of the circular column is 17% higher than that of the square one. However, the ultimate pure bending moment of the circular column is only 0.6% higher than that of the square one. This indicates that circular slender CFSST columns perform much better than square ones when subjected to axial compression. However, for members under bending, the shape effect is minor.

4.7.8 Column strength curves

Column R9 presented in Table 4.2 was analyzed by means of varying its column slenderness ratio to investigate the influences of the loading eccentricity ratio (e/D) on the column strength curves for square CFSST columns. The column strength curves obtained are shown in Figure 4.19, where P_{oa} represents the ultimate axial strength of the column cross-section with an eccentricity ratio of 0.1. It can be seen from the figure that increasing the column slenderness ratio (L/r) causes a significant reduction in the column ultimate axial strength regardless of the loading eccentricity ratio (e/D). The loading eccentricity has the most pronounced effect on the ultimate axial strength of the cross-sections, and its effect diminishes as the column slenderness ratio increases. When the e/D ratio is increased from 0.1 to 0.2, 0.4, and 0.6, the cross-section strength decreases by 20%, 48%, and 63%, respectively. For a very long beam-column having an L/r ratio of 200, its ultimate axial strength is reduced by only 13%, 28%, and 38%, respectively, when increasing the e/D ratio from 0.1 to 0.2, 0.4, and 0.6.

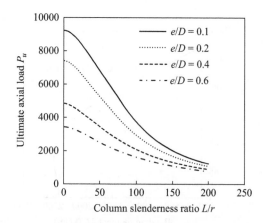

Figure 4.19   Effects of the eccentricity ratio on the column strength curves of square CFSST beam-columns.

4.8 DESIGN OF RECTANGULAR AND SQUARE CFSST SLENDER COLUMNS

4.8.1 Ultimate pure moments of square columns

Liang and Fragomeni (2010) proposed an equation for calculating the ultimate pure moment capacity (M_o) of circular CFST columns. This equation is adopted herein for square CFSST columns considering the local buckling of the stainless steel tube and is expressed by

$$M_o = \lambda_m \alpha_c \alpha_s Z_e \sigma_{0.2} \tag{4.19}$$

where Z_e denotes the elastic section modulus of the square CFSST column, computed by $Z_e = \pi D^3/32$. The factors λ_m, α_c, and α_s account for the effects of the depth-to-thickness ratio D/t, concrete strength f_c', and stainless steel proof stress $\sigma_{0.2}$, respectively.

The effects of local buckling and strain-hardening of stainless steel have been taken into account in the numerical analyses of CFSST columns under pure bending, and the results obtained have been used to derive equations for determining the factors λ_m, α_c, and α_s (Patel et al. 2017). The factor λ_m incorporates the effect of the D/t ratio on the ultimate pure moment (M_o) and is expressed by

$$\lambda_m = 0.0187 + 24.2\left(\frac{D}{t}\right)^{-1} - 61.1\left(\frac{D}{t}\right)^{-2} \qquad \text{for } 10 \le \frac{D}{t} \le 120 \tag{4.20}$$

The factor α_c is used to consider the effect of the concrete compressive strength f_c' on the ultimate pure moment capacity of CFSST columns. The equation for predicting the factor α_c is given as

$$\alpha_c = 0.7266\left(f_c'\right)^{0.0668} \qquad \text{for } 20 \le f_c' \le 120 \text{ MPa} \tag{4.21}$$

The factor α_s reflects the influence of the stainless steel proof stress $\sigma_{0.2}$ on M_o and is computed by

$$\alpha_s = 0.471 + \frac{262.62}{\sigma_{0.2}} - \frac{28119}{\sigma_{0.2}^2} \qquad \text{for } 250 \le \sigma_{0.2} \le 690 \text{ MPa} \tag{4.22}$$

The ultimate pure moments calculated using Eq. (4.19) are compared with the fiber element analysis predictions in Table 4.3. The CFSST columns with various stainless steel proof stresses, concrete strengths, and depth-to-thickness ratios are considered in the verification of the proposed equation. In Table 4.3, $M_{o.cal}$ stands for the pure moment capacity obtained from

Table 4.3 Comparison of ultimate pure moments of square CFSST columns

Specimens	$B \times D \times t$ (mm)	D/t	f'_c (MPa)	$\sigma_{0.2}$ (MPa)	E_0 (GPa)	n	$M_{o.fib}$ (kNm)	$M_{o.cal}$ (kNm)	$\dfrac{M_{o.cal}}{M_{o.fib}}$
S1	$300 \times 300 \times 6$	50	65	250	200	7	307	327	1.06
S2	$300 \times 300 \times 4.29$	70	65	250	200	7	247	241	0.97
S3	$300 \times 300 \times 3.75$	80	65	250	200	7	220	213	0.97
S4	$300 \times 300 \times 3.33$	90	65	250	200	7	199	191	0.96
S5	$300 \times 300 \times 3.00$	100	65	250	200	7	182	174	0.96
S6	$300 \times 300 \times 5$	60	20	250	200	7	259	256	0.99
S7	$300 \times 300 \times 5$	60	25	250	200	7	262	260	0.99
S8	$300 \times 300 \times 5$	60	32	250	200	7	266	264	0.99
S9	$300 \times 300 \times 5$	60	65	250	200	7	282	277	0.98
S10	$300 \times 300 \times 5$	60	65	250	200	7	288	281	0.97
S11	$300 \times 300 \times 5$	60	65	250	200	7	295	285	0.97
S12	$500 \times 500 \times 10$	50	100	250	200	7	1590	1556	0.98
S13	$500 \times 500 \times 10$	50	100	350	200	7	2037	2017	0.99
S14	$500 \times 500 \times 10$	50	100	550	200	7	2721	2734	1.00
Mean									0.98
Standard deviation (SD)									0.025
Coefficient of variation (COV)									0.025

Eq. (4.19) and $M_{o.fib}$ represents the pure moment capacity obtained from the fiber element model. The ratio of the calculated pure moment capacity to the fiber analysis ultimate pure moment $\left(M_{o.cal}/M_{o.fib}\right)$ varies from 0.96 to 1.00, and the corresponding mean, coefficient of variation, and standard deviation are 0.98, 0.025, and 0.025, respectively. This shows that the proposed design equation provides good predictions of the pure moment capacity M_o of square CFSST columns.

4.8.2 Slender columns under axial compression

Eurocode 4 (2004) provides design equations for concentrically loaded slender CFST columns, which can be utilized to calculate the member capacity of square slender CFSST columns. The design equations given in Eurocode 4 are based on the buckling curves which account for the influences of column buckling and axial load–moment interaction. The influences of the plastic behavior of steel and cracking of concrete are also considered. The design equation given in Eurocode 4 is limited to the column relative slenderness ratio of $\bar{\lambda} \leq 2.0$. The member capacity of concentrically loaded rectangular slender CFSST columns is calculated by

$$P_{u.EC4} = \chi N_{pl.Rd} \qquad (4.23)$$

where $N_{pl.Rd}$ represents the axial resistance of the rectangular cross-section, which is determined as

$$N_{pl.Rd} = A_c f_c' + A_s \sigma_{0.2} \tag{4.24}$$

In Eq. (4.23), χ denotes the reduction factor for the buckling curve "a" and is expressed by

$$\chi = \frac{1}{\phi + \sqrt{\phi^2 - \bar{\lambda}^2}} \tag{4.25}$$

where

$$\phi = 0.5\left[1 + 0.21\left(\bar{\lambda} - 0.2\right) + \bar{\lambda}^2\right] \tag{4.26}$$

The relative slenderness ratio $\bar{\lambda}$ is defined as

$$\bar{\lambda} = \sqrt{\frac{N_{pl.Rd}}{N_{cr}}} \tag{4.27}$$

where N_{cr} represents the elastic buckling load of the column, which is calculated by the following formula:

$$N_{cr} = \frac{(EI)_{eff}\, \pi^2}{L^2} \tag{4.28}$$

in which $(EI)_{eff}$ is the effective flexural stiffness of the rectangular CFSST column, which is computed by

$$(EI)_{eff} = E_0 I_s + 0.6 E_{cm} I_c \tag{4.29}$$

where I_s and I_c stand for the second moment of area of the stainless steel tube and filled concrete, respectively. The factor of 0.6 in Eq. (4.29) accounts for the concrete cracking caused by the bending moment due to second-order effects.

In Eq. (4.29), E_{cm} is the modulus of concrete, which is computed by

$$E_{cm} = 22,000\left(\frac{f_c' + 8}{10}\right)^{0.3} \tag{4.30}$$

The ultimate axial strengths of the test specimens are compared with those computed by Eq. (4.23) in Table 4.1. The mean computed-to-tested strength ratio was 0.91 with the coefficient of variation and standard deviation being 0.079 and 0.089, respectively. It can be concluded that the design method given in Eurocode 4 (2004) gives conservative estimates of the ultimate axial strengths of rectangular slender CFSST columns under axial compression. Therefore, the design equation given in Eurocode 4 for rectangular CFST columns can be used to calculate the ultimate strength of axially loaded rectangular CFSST columns.

4.9 CONCLUSIONS

This chapter has presented theoretical models formulated on the basics of the fiber element method for the predictions of axial load–deflection responses and axial load–moment strength envelopes of rectangular slender CFSST beam-columns under axial loading and biaxial bending. The theoretical models have accounted for the effects of different strain-hardening behaviors of stainless steel in compression and tension and the interaction of local and overall buckling on the fundamental behavior of biaxially loaded rectangular slender CFSST beam-columns. Computer simulation procedures and solution algorithms for the determination of axial load–deflection curves and strength envelopes have been described in detail. The theoretical models have been verified by comparisons with experimental data. Parametric studies have been conducted by employing the computer program that incorporates the theoretical models to investigate the effects of material and geometric parameters on the behavior of slender CFSST beam-columns. A simple equation for predicting the ultimate pure moment of square CFSST columns has been proposed. It has been shown that the design equation given in Eurocode 4 yields conservative predictions of the ultimate axial strength of rectangular CFSST columns.

The following conclusions have been drawn from the numerical results:

- The column ultimate axial strength is significantly reduced by increasing the depth-to-thickness ratio, loading eccentricity ratio, and column slenderness ratio but is significantly increased by increasing the material strengths of either stainless steel or concrete.
- The concrete contribution to the ultimate axial load of a CFSST column is reduced by using a larger loading eccentricity or column slenderness but is increased by using higher concrete strength.
- The ultimate pure moment capacity of a CFSST column depends on the depth-to-thickness ratio, concrete compressive strength, and stainless steel yield strength.
- The column initial stiffness is notably reduced by increasing either the depth-to-thickness ratio or the concrete strength but is remarkably reduced by increasing either the loading eccentricity ratio or the column slenderness ratio.

- The column displacement–ductility generally decreases with increasing the material strengths but increases as the geometric parameters increases.
- The local buckling of stainless steel tubes considerably reduces the stiffness, strength, and ductility of the CFSST beam-columns. It has the most pronounced effect on the ultimate axial strength of the column cross-sections, but its effect decreases as the column slenderness increases.
- The applied load angle does not have a significant influence on the ultimate strength of the CFSST slender columns.
- For circular and square CFSST columns with the same cross-sectional area, e/D ratio, L/r ratio, and material yield strengths, the circular column performs better than the square one when subjected to axial compression and bending, but this effect is minor when subjected to pure bending.
- The loading eccentricity has the most pronounced influence on the ultimate axial strength of cross-sections, and its effect diminishes as the column slenderness increases.

REFERENCES

Bridge, R. Q. (1976) "Concrete filled steel tubular columns," Research Report No. R 283, School of Civil Engineering, The University of Sydney, Sydney, Australia.

Eurocode 4 (2004) *Design of Composite Steel and Concrete Structures, Part 1.1: General Rules and Rules for Building*, Brussels, Belgium: European Committee for Standardization, CEN.

Guo, L., Zhang, S. and Xu, Z. (2011) "Behaviour of filled rectangular steel HSS composite columns under bi-axial bending," *Advances in Structural Engineering*, 14(2): 295–306.

Lai, Z., Varma, A. H. and Zhang, K. (2014) "Noncompact and slender rectangular CFT members: Experimental database, analysis and design," *Journal of Constructional Steel Research*, 101: 455–468.

Lakshmi, B. and Shanmugam, N. E. (2002) "Nonlinear analysis of in-filled steel-concrete composite columns," *Journal of Structural Engineering*, ASCE, 128(7): 922–933.

Liang, Q. Q. (2009a) "Performance-based analysis of concrete-filled steel tubular beam-columns, Part I: Theory and algorithms," *Journal of Constructional Steel Research*, 65(2): 363–372.

Liang, Q. Q. (2009b) "Performance-based analysis of concrete-filled steel tubular beam-columns, Part II: Verification and applications," *Journal of Constructional Steel Research*, 65(2): 351–362.

Liang, Q. Q. (2011a) "High strength circular concrete-filled steel tubular slender beam-columns, Part I: Numerical analysis," *Journal of Constructional Steel Research*, 67(2): 164–171.

Liang, Q. Q. (2011b) "High strength circular concrete-filled steel tubular slender beam-columns, Part II: Fundamental behavior," *Journal of Constructional Steel Research*, 67(2): 172–180.

Liang, Q. Q. (2018) "Numerical simulation of high strength circular double-skin concrete-filled steel tubular slender columns," *Engineering Structures*, 168: 205–217.

Liang, Q. Q. and Fragomeni, S. (2010) "Nonlinear analysis of circular concrete-filled steel tubular short columns under eccentric loading," *Journal of Constructional Steel Research*, 66(2): 159–169.

Liang, Q. Q., Patel, V. I. and Hadi, M. N. S. (2012) "Biaxially loaded high-strength concrete-filled steel tubular slender beam-columns, Part I: Multiscale simulation," *Journal of Constructional Steel Research*, 75: 64–71.

Liang, Q. Q., Uy, B. and Liew, J. Y. R. (2007) "Local buckling of steel plates in concrete-filled thin-walled steel tubular beam-columns," *Journal of Constructional Steel Research*, 63(3): 396–405.

Liew, J. Y. R., Xiong, M. X. and Xiong, D. X. (2016) "Design of concrete filled tubular beam-columns with high strength steel and concrete," *Structures*, 8: 213–226.

Müller, D. E. (1956) "A method for solving algebraic equations using an automatic computer," *MTAC*, 10: 208–215.

Mursi, M. and Uy, B. (2004) "Strength of slender concrete filled high strength steel box columns," *Journal of Constructional Steel Research*, 60(12): 1825–1848.

Mursi, M. and Uy, B. (2006) "Behaviour and design of fabricated high strength steel columns subjected to biaxial bending Part II: analysis and design codes," *Advanced Steel Construction*, 2(4): 316–354.

Patel, V. I., Liang Q. Q. and Hadi M. N. S. (2012) "High strength thin-walled rectangular concrete-filled steel tubular slender beam-columns, Part I: Modeling," *Journal of Constructional Steel Research*, 70: 377–384.

Patel, V. I., Liang, Q. Q. and Hadi, M. N. S. (2015a) "Biaxially loaded high-strength concrete-filled steel tubular slender beam-columns, Part II: Parametric Study," *Journal of Constructional Steel Research*, 110: 200–207.

Patel, V. I., Liang, Q. Q. and Hadi, M. N. S. (2015b) *Nonlinear Analysis of Concrete-Filled Steel Tubular Columns*, Germany: Scholar's Press.

Patel, V. I., Liang, Q. Q. and Hadi, M. N. S. (2017) "Nonlinear analysis of biaxially loaded rectangular concrete-filled stainless steel tubular slender beam-columns," *Engineering Structures*, 140: 120–133.

Shakir-Khalil, H. and Zeghiche, J. (1989) "Experimental behaviour of concrete-filled rolled rectangular hollow-section columns," *The Structural Engineer*, 67(19): 346–353.

Tokgoz, S. (2015) "Tests on plain and steel fiber concrete-filled stainless steel tubular columns," *Journal of Constructional Steel Research*, 114: 129–135.

Uy, B., Tao, Z. and Han, L. H. (2011) "Behaviour of short and slender concrete-filled stainless steel tubular columns," *Journal of Constructional Steel Research*, 67(3): 360–378.

Wang, Y. C. (1999) "Tests on slender composite columns," *Journal of Constructional Steel Research*, 49(1): 25–41.

Xiong, M. X., Xiong, D. X. and Liew, J. Y. R. (2017) "Behaviour of steel tubular members infilled with ultra high strength concrete," *Journal of Constructional Steel Research*, 138: 168–183.

Notations

a_m	Coefficient of Müller's method
A_c	Cross-sectional area of concrete
$A_{c,j}$	Area of the jth concrete fiber element
A_s	Cross-sectional area of the stainless steel tube
$A_{s,i}$	Area of the ith steel fiber element
b	Clear width of a stainless steel plate
b_{e1}, b_{e2}	Effective widths of a stainless steel plate
b_m	Coefficient of Müller's method
b_{ne}	Ineffective width of a stainless steel plate
$b_{ne,max}$	Maximum value of ineffective width for a stainless steel plate
B	Width of a rectangular cross-section
B_s	Width taken as the larger value of B and D for a rectangular cross-section
C_1, C_2, C_3, C_4	Material constants of stainless steel
C_5, C_6, C_7, C_8	Material constants of stainless steel
$d_{e,i}$	Orthogonal distance from the centroid of each fiber element to the neutral axis in a composite column cross-section
d_n	Depth of neutral axis in a composite column section
D	Depth of a rectangular cross-section or diameter of a circular cross-section
D_c	Diameter of concrete core
e	Loading eccentricity
e_x	Loading eccentricity along x-axis
e_y	Loading eccentricity along y-axis
E_0	Young's modulus of stainless steel
$E_{0.2}$	Tangent modulus of stainless steel

E_c, E_{cm}	Young's modulus of concrete
$(EI)_{eff}$	Effective flexural stiffness of a composite column
f'_c	Compressive strength of concrete cylinder
f'_{cc}	Compressive strength of confined concrete
f_{ct}	Tensile strength of concrete
f_{rp}	Lateral confining pressure provided by a circular steel tube on concrete
f_y	Yield strength of carbon steel
f_u	Tensile strength of structural steel
I_c	Second moment of area of concrete in a composite section
I_s	Second moment of area of steel section in composite section
L	Length of column
m	Strain-hardening factor of stainless steel
M_e	Moment at the column ends
$M_{e.max}$	Maximum moment at the ends of a column
M_{me}	External bending moment at the mid-height of a slender column
M_{mi}	Resultant bending moment at the mid-height of a column
M_o	Ultimate pure moment capacity of a column
$M_{o.cal}$	Ultimate pure bending strength of a square composite section calculated by the design equation
$M_{o.fib}$	Ultimate pure bending strength of a square composite section predicted by the fiber model
M_u	Ultimate moment capacity of a column under combined axial load and bending
M_x	Section moment capacity of a short composite beam-column bending about the major principal x-axis
M_y	Section moment capacity of a short composite beam-column bending about the minor principal y-axis
n	Nonlinear index of stainless steel
$n'_{0.2,1.0}$	Strain-hardening parameter of stainless steel
n_2	Material constant of stainless steel
nc	Total number of concrete fiber element
ns	Total number of steel fiber element
N_{cr}	Elastic buckling load of a column
$N_{pl.Rd}$	Ultimate axial strength of composite column cross-section
$N_{pl.Rk}$	Plastic resistance of a composite column under axial compression
P	Axial force

P_a	Applied axial load
P_{max}	Maximum axial load of a column
P_o	Ultimate axial load of a short composite column under pure axial load
P_{oa}	Ultimate axial load of a slender composite column under pure axial load
P_{oe}	Ultimate strength of eccentrically loaded composite cross-section
P_s	Ultimate axial strength of a hollow stainless steel tubular beam-column
P_u	Ultimate axial load of a short composite column under axial compression
ΔP_u	Axial load increment
$P_{u.c}$	Axial load carried by the concrete core when CFSST column attains its ultimate strength
$P_{u.exp}$	Ultimate axial strength of a short composite column obtained from experiments
$P_{u.fib}$	Ultimate axial strength of a short composite column predicted by the fiber model
$P_{u.s}$	Axial load carried by the stainless steel tube when CFSST column attains its ultimate strength
$P_{u.AISC}$	Ultimate axial strength of a circular short composite column predicted using AISC 316-16 design code
$P_{u.EC4}$	Ultimate axial strength of a circular short composite column predicted using Eurocode 4
$P_{u.Patel}$	Ultimate axial strength of a circular short composite column predicted using Patel et al. model
PI_{ad}	Strain ductility of a short composite column under axial compression
PI_{cd}	Curvature ductility of a short composite column under axial compression
r	Radius of gyration of a cross-section
r_m, r_{pu}	Residual moment in a composite column section
r_p	Residual force in a composite column section
R_c, ξ_c	Concrete contribution ratio
R_s	Stainless steel contribution ratio
t	Thickness of a steel tube
u	Displacement
u_m	Lateral deflection at the mid-height of column
Δu_m	Deflection increment at the mid-height of a column
u_o	Initial geometric imperfection at the mid-height of a slender column

v_e	Poisson's ratio of the steel tube with concrete infill
v_s	Poisson's ratio of the steel tube without concrete infill
x_i	Centroid x coordinate of fiber element i
$x_{n,i}$	Distance from the centroid of the ith fiber element
y_i	Centroid y coordinate of fiber element i
$y_{n,i}$	Distance from the centroid of the ith fiber element
Z_e	Elastic section modulus
α	Applied load angle with respect to the y-axis of a composite column section
α_s	Stress gradient coefficient
β_c	Strength degradation factor of concrete in rectangular steel tube
β_{cc}	Strength degradation factor of concrete in circular steel tube
χ	Reduction factor for slender columns under axial compression used in Eurocode 4
$\varepsilon_{0.2}$	0.2% proof strain of stainless steel
$\varepsilon_{1.0}$	1.0% proof strain of stainless steel
$\varepsilon_{2.0}$	2.0% proof strain of stainless steel
ε_c	Concrete strain
ε_c'	Unconfined concrete compressive strain at f_c'
ε_{cc}'	Confined concrete compressive strain at f_{cc}'
ε_{cu}	Ultimate concrete compressive strain
ε_i	Strain at the ith fiber element
ε_k	Convergence tolerance
ε_{max}	Maximum axial strain of a composite column
ε_r	Strain ratio $\varepsilon_{ss}/\varepsilon_{0.2}$ of stainless steel
ε_s	Strain in a steel fiber
ε_{ss}	Strain in a stainless steel fiber
ε_{su}	Ultimate strain of steel
ε_{ssu}	Ultimate strain of stainless steel
ε_t	Steel strain at strain hardening or maximum strain at top fiber
ε_{tc}	Concrete cracking strain at f_{ct}
ε_{tu}	Ultimate tensile strain of concrete
ε_{ut}	Ultimate tensile strain of stainless steel
ε_y	Yield strain of structural steel
$\varepsilon_{y.75}$	Strain at 75% section capacity in the ascending curve of the load–strain response

$\varepsilon_{u.90}$	Strain at 90% section capacity in the descending curve of the load–strain response
ϕ	Curvature
$\phi_{0.75}$	Curvature at the 75% bending strength in the ascending branch of moment–curvature response
ϕ_e	Curvature at the column ends
ϕ_m	Curvature at the mid-height of a slender composite beam-column
$\Delta\phi_m$	Curvature increment
ϕ_u	Curvature when the moment falls to 90% bending strength in the descending branch of moment–curvature response
γ_c	Strength reduction factor of concrete
γ_s	Strength factor accounting for the effect of hoop tensile stresses and strain-hardening on the yield stress of the steel tube
η_a, η_c	Concrete confinement factors given in Eurocode 4
$\overline{\lambda}$	Relative slenderness of a composite column
θ	Orientation of the neutral axis with respect to the x-axis in a composite column section
$\sigma_{0.01}$	0.01% proof stress of stainless steel
$\sigma_{0.2}$	0.2% proof stress of stainless steel
σ_1	Maximum edge stress on a plate
$\sigma_{1.0}$	1.0% proof stress of stainless steel
σ_2	Minimum edge stress on a plate
$\sigma_{2.0}$	2.0% proof stress of stainless steel
σ_{lc}	Initial local buckling stress of a clamped steel plate
σ_{lu}	Ultimate value of the maximum edge stress σ_1 on a plate
σ_c	Compressive concrete stress
$\sigma_{c,j}$	Compressive concrete stress in a concrete fiber element j
σ_s	Steel stress
$\sigma_{s,i}$	Stress in a steel fiber element i
σ_{ss}	Stress in a stainless steel fiber
σ_{ssu}	Ultimate strength of stainless steel
σ_{ut}	Ultimate tensile strength of stainless steel
$\omega_1, \omega_2, \omega_3, \omega_4$	Design variables

Index

131